The War Machines

FIGHTING BY SEA AND SKY

The War Machines
The Development of Aircraft and Submarines for Aerial and Undersea Warfare to the End of the First World War

Willis J. Abbot

The War Machines: the Development of Aircraft and Submarines for Aerial and Undersea Warfare to the End of the First World War
by Willis J. Abbot

Originally published in 1918 under the title
Aircraft & Submarines: the Story of the Invention, Development and Present-Day Uses of War's Newest Weapons

Leonaur is an imprint of Oakpast Ltd

Material original to this edition and presentation of text in this form copyright © 2009 Oakpast Ltd

ISBN: 978-0-85706-126-3 (hardcover)
ISBN: 978-0-85706-125-6 (softcover)

http://www.leonaur.com

Publisher's Notes

The views expressed in this book are not necessarily those of the publisher.

Contents

Preface	7

The Conquest of the Air

Introductory	11
The Earliest Flying Men	22
The Services of Santos-Dumont	44
The Count von Zeppelin	61
The Development of the Aeroplane	82
The Training of the Aviator	104
Some Methods of the War in the Air	122
Incidents of the War in the Air	157
The United States at War	180
Some Features of Aerial Warfare	203

The Submarine Boat

Beginnings of Submarine Invention	229
The Coming of Steam and Electricity	249
John P. Holland and Simon Lake	264
The Modern Submarine	285
Aboard a Submarine	305
Submarine Warfare	319
The Future of the Submarin	350

Preface

Not since gunpowder was first employed in warfare has so revolutionary a contribution to the science of slaughtering men been made as by the perfection of aircraft and submarines. The former have had their first employment in this world-wide war of the nations. The latter, though in the experimental stage as far back as the American Revolution, have in this bitter contest been for the first time brought to so practical a stage of development as to exert a really appreciable influence on the outcome of the struggle.

Comparatively few people appreciate how the thought of navigating the air's dizziest heights and the sea's gloomiest depths has obsessed the minds of inventors. From the earliest days of history men have grappled with the problem, yet it is only within two hundred years for aircraft and one hundred for submarines that any really intelligent start has been made upon its solution. The men who really gave practical effect to the vague theories which others set up—in aircraft the Wrights, Santos-Dumont, and Count Zeppelin; in submarines Lake and Holland—are either still living, or have died so recently that their memory is still fresh in the minds of all.

In this book the author has sketched swiftly the slow stages by which in each of these fields of activity success has been attained. He has collated from the immense mass of records of the activities of both submarines and aircraft enough interesting data to show the degree of perfection and practicability to which both have been brought. And he has outlined so far as possible from existing conditions the possibilities of future usefulness in fields other than those of war of these new devices.

The most serious difficulty encountered in dealing with the present state and future development of aircraft is the rapidity with which that development proceeds. Before a Congressional Committee last January an official testified that grave delay in the manufacture of aeroplanes for the army had been caused by the fact that types adopted a scant three months before had become obsolete, because of experience on the European battlefields, and later inventions before the first machines could be completed. There may be exaggeration in the statement but it is largely true. Neither the machines nor the tactics employed at the beginning of the war were in use in its fourth year. The course of this evolution, with its reasons, are described in this volume.

Opportunities for the peaceful use of aeroplanes are beginning to suggest themselves daily. After the main body of this book was in type the Postmaster-General of the United States called for bids for an aerial mail service between New York and Washington—an act urged upon the Government in this volume. That service contemplates a swift carriage of first-class mail at an enhanced price—the tentative schedule being three hours, and a postage fee of twenty-five cents an ounce. There can be no doubt of the success of the service, its value to the public, and its possibilities of revenue to the post-office. Once its usefulness is established it will be extended to routes of similar length, such as New York and Boston, New York and Buffalo, or New York and Pittsburgh. The mind suggests no limit to the extension of aerial service, both postal and passenger, in the years of industrial activity that shall follow the war.

In the preparation of this book the author has made use of many records of personal experiences of those who have dared the air's high altitudes and the sea's stilly depths. For permission to use certain of these he wishes to express his thanks to the Century Co., for extracts from *My Airships* by Santos-Dumont; to Doubleday, Page & Co., for extracts from *Flying for France*, by James R. McConnell; to Charles Scribner's Sons, for material drawn from *With the French Flying Corps*, by Carroll Dana Winslow; to *Collier's Weekly*, for certain extracts from interviews with Wilbur Wright; to *McClure's Magazine*, for the account of Mr. Ray Stannard Baker's trip in a Lake submarine; to Hearst's International Library, and to the *Scientific American*, for the use of several illustrations.

W. J. A.
New York, 1918

The Conquest of the Air

CHAPTER 1

Introductory

It was at Mons in the third week of the Great War. The grey-green German hordes had overwhelmed the greater part of Belgium and were sweeping down into France whose people and military establishment were all unprepared for attack from that quarter. For days the little British army of perhaps 100,000 men, that forlorn hope which the Germans scornfully called "contemptible," but which man for man probably numbered more veteran fighters than any similar unit on either side, had been stoutly holding back the enemy's right wing and fighting for the delay that alone could save Paris. At Mons they had halted, hoping that here was the spot to administer to von Kluck, beating upon their front, the final check. The hope was futile. Looking back upon the day with knowledge of what General French's army faced—a knowledge largely denied to him—it seems that the British escape from annihilation was miraculous. And indeed it was due to a modern miracle—the conquest of the air by man in the development of the aeroplane.

General French was outnumbered and in danger of being flanked on his left flank. His right he thought safe, for it was in contact with the French line which extended eastward along the bank of the Somme to where the dark fortress of Namur frowned on the steeps formed by the junction of that river with the Meuse. At that point the French line bent to the south following the course of the latter river.

Namur was expected to hold out for weeks. Its defence lasted but three days! As a matter of fact it did not delay the oncoming Germans a day, for they invested it and drove past in their fierce assault upon Joffre's lines. Enormously outnumbered, the French were broken and forced to retreat. They left General French's right flank in the air, exposed to envelopment by von Kluck who was already reaching around the left flank. The German troops were ample in number to surround the British, cut them off from all support, and crush or capture them all. This indeed they were preparing to do while General French, owing to some mischance never yet explained, was holding his ground utterly without knowledge that his allies had already retired leaving his flank without protection.

When that fatal information arrived belatedly at the British headquarters it seemed like a death warrant. The right of the line had already been exposed for more than half-a-day. It was inexplicable that it had not already been attacked. It was unbelievable that the attack would not fall the next moment. But how would it be delivered and where, and what force would the enemy bring to it? Was von Kluck lulling the British into a false sense of security by leaving the exposed flank unmenaced while he gained their rear and cut off their retreat? Questions such as these demanded immediate answer. Ten years before the most dashing scouts would have clattered off to the front and would have required a day, perhaps more, to complete the necessary reconnaissance. But though of all nations, except of course the utterly negligent United States, Great Britain had least developed her aviation corps, there were attached to General French's headquarters enough airmen to meet this need. In a few minutes after the disquieting news arrived the beat of the propellers rose above the din of the battlefield and the aeroplanes appeared above the enemy's lines. An hour or two sufficed to gather the necessary facts, the fliers returned to headquarters, and immediately the retreat was begun.

It was a beaten army that plodded back to the line of the Marne. Its retreat at times narrowly approached a rout. But the army was not crushed, annihilated. It remained a coherent, serviceable part of the allied line in the successful action speedily fought along the Marne. But had it not been for the presence of the airmen the British expeditionary force would have been wiped out then and there.

The battle of Mons gave the soldiers a legend which still persists— that of the ghostly English bowmen of the time of Edward the Black Prince who came back from their graves to save that field for England and for France. Thousands of simple souls believe that legend to-day. But it is no whit more unbelievable than the story of an army saved by a handful of men flying thousands of feet above the field would have been had it been told of a battle in our Civil War. The world has believed in ghosts for centuries and the Archers of Mons are the legitimate successors of the Great Twin Brethren at the Battle of Lake Regillus. But

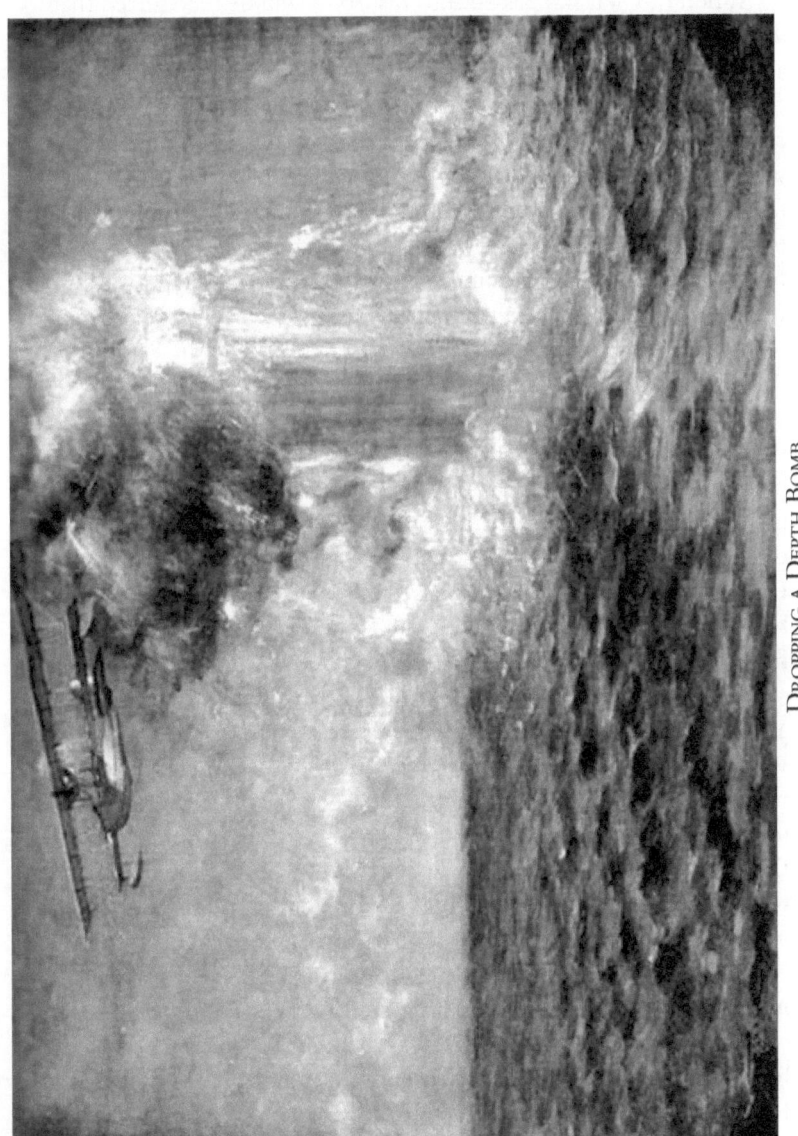

Dropping a Depth Bomb

Caesar, Napoleon, perhaps the elder von Moltke himself would have scoffed at the idea that men could turn themselves into birds to spy out the enemy's dispositions and save a sorely menaced army.

When this war has passed into history it will be recognized that its greatest contributions to military science have been the development and the use of aircraft and submarines. There have, of course, been other features in the method of waging war which have been novel either in themselves, or in the gigantic scale upon which they have been employed. There is, for example, nothing new about trench warfare. The American who desires to satisfy himself about that need only to visit the Military Park at Vicksburg, or the country about Petersburg or Richmond, to recognize that even fifty years ago our soldiers understood the art of sheltering themselves from bullet and shrapnel in the bosom of Mother Earth. The trench warfare in Flanders, the Argonne, and around Verdun has been novel only in the degree to which it has been developed and perfected. Concrete-lined trenches, with spacious and well-furnished bomb-proofs, with phonographs, printing presses, and occasional dramatic performances for lightening the soldiers' lot present an impressive elaboration of the muddy ditches of Virginia and Mississippi. Yet after all the boys of Grant and Lee had the essentials of trench warfare well in mind half a century before Germany, France, and England came to grips on the long line from the North Sea to the Vosges.

Asphyxiating gas, whether liberated from a shell, or released along a trench front to roll slowly down before a wind upon its defenders, was a novelty of this war. But in some degree it was merely a development of the "stinkpot" which the Chinese have employed for years. So too the tear-bomb, or lachrymatory bomb, which painfully irritated the eyes of all in its neighbourhood when it burst, filling them with tears and making the soldiers practically helpless in the presence of a swift attack. These two weapons of offence, and particularly the first, because of the frightful and long-continuing agony it inflicts upon its victims, fascinated the observer, and awakened the bitter protests of those who held that an issue at war might be determined by civilized nations without recourse to engines of death and anguish more barbaric than any known to the red Indians, or the most savage tribes of Asia. Neither of these devices, nor for that matter the cognate one of fire spurted like a liquid from a hose upon a shrinking enemy, can be shown to have had any appreciable effect upon the fortunes of any great battle. Each, as soon as employed by any one belligerent, was quickly seized by the adversary, and the respiratory mask followed fast upon the appearance of the chlorine gas. Whatever the outcome of the gigantic conflict may be, no one will claim that any of these devices had contributed greatly to the result.

But the aeroplane revolutionized warfare on land. The submarine has made an almost equal revolution in naval warfare.

Had the aeroplane been known in the days of our Civil War some of its most picturesque figures would have never risen to eminence or at least would have had to win their places in history by efforts of an entirely different sort. There is no place left in modern military tactics for the dashing cavalry scout of the type of Sheridan, Custer, Fitz Lee, or Forrest. The aeroplane, soaring high above the lines of the enemy, brings back to headquarters in a few hours information that in the old times took a detachment of cavalry days to gather. The "screen of cavalry" that in bygone campaigns commanders used to mask their movements no longer screens nor masks. A general moves with perfect knowledge that his enemy's aircraft will report to their headquarters his roads, his strength, and his probable destination as soon as his vanguard is off. During the Federal advance upon Richmond, Stonewall Jackson, most brilliant of the generals of that war, repeatedly slipped away from the Federal front, away from the spot where the Federal commanders confidently supposed him to be, and was found days later in the Valley of the Shenandoah, threatening Washington or menacing the Union rear and its communications. The war was definitely prolonged by this Confederate dash and elusiveness—none of which would have been possible had the Union forces possessed an aviation corps.

It is yet to be shown conclusively that as offensive engines aircraft have any great value. The tendency of the military authorities of every side to minimize the damage they have suffered makes any positive conclusion on this subject difficult and dangerous at this moment. The aeroplane by day or the Zeppelin by night appears swiftly and mysteriously, drops its bombs from a height of several thousand feet, and takes its certain flight through the boundless sky to safety. The aggressor cannot tell whether his bombs have found a fitting target. He reports flaming buildings left behind him, but whether they are munitions factories, theatres, or primary schools filled with little children he cannot tell. Nor does he know how quickly the flames were extinguished, or the amount of damage done. The British boast of successful air raids upon Cuxhaven, Zeebrugge, Essen, and Friedrich-

A Battle in Mid-Air: Note Rifleman on Wing of Aeroplane

shaven. But if we take German official reports we must be convinced that the damage done was negligible in its relation to the progress of the war. In their turn the Germans brag mightily of the deeds of their Zeppelins over London, and smaller British towns. But the sum and substance of their accomplishment, according to the British reports, has been the slaughter and mutilation of a number of civilians—mostly women and children—and the bloody destruction of many humble working-class homes.

At this writing, December, 1917, it is not recorded that any battleship, munitions factory, any headquarters, great government building, or fortress has been destroyed or seriously injured by the activities of aircraft of either type. This lack of precise information may be due to the censor rather than to any lack of great deeds on the part of airmen. We do know of successful attacks on submarines, though the military authorities are chary about giving out the facts. But as scouts, messengers, and guides for hidden batteries attacking unseen targets, aviators have compelled the rewriting of the rules of military strategy. About this time, however, it became apparent that the belligerents intended to develop the battle planes. Particularly was this true of the Allies. The great measure of success won by the German submarines and the apparent impossibility of coping adequately with those weapons of death once they had reached the open sea, led the British and the Americans to consider the possibility of destroying them in their bases and destroying the bases as well. But Kiel and Wilhelmshaven were too heavily defended to make an attack by sea seem at all practicable. The lesser ports of Zeebrugge and Ostend had been successfully raided from the air and made practically useless as submarine bases. Discussion therefore was strong of making like raids with heavier machines carrying heavier guns and dropping more destructive bombs upon the two chief lurking places of the submarines. While no conclusion had been reached as to this strategy at the time of the publication of this book, both nations were busy building larger aircraft probably for use in such an attack.

*****

The submarine has exerted upon the progress of the war an influence even more dominant than that of aircraft. It has been a positive force both offensive and defensive. It has been Germany's only potent weapon for bringing home to the British the privations and want which war entails upon a civilian population, and at the same time guarding the German people from the fullest result of the British blockade. It is no overstatement to declare that but for the German submarines the war would have ended in the victory of the Allies in 1916.

We may hark back to our own Civil War for an illustration of the crushing power of a superior navy not qualified by any serviceable weapon in the hands of the weaker power.

Historians have very generally failed to ascribe to the Federal blockade of Confederate ports its proportionate influence on the outcome of that war. The Confederates had no navy. Their few naval vessels were mere commerce destroyers, fleeing the ships of the United States navy and preying upon unarmed merchantmen. With what was rapidly developed into the most powerful navy the world had ever seen, the United States Government from the very beginning of the war locked the Confederate States in a wall of iron. None might pass going in or out, except by stealth and at the peril of property and life. Outside the harbour of every seaport in the control of the Confederates the blockading men-of-war lurked awaiting the blockade runners. Their vigilance was often eluded, of course, yet nevertheless the number of cargoes that slipped through was painfully inadequate to meet the needs of the fenced-in States. Clothing, medicines, articles of necessary household use were denied to civilians. Cannon, rifles, saltpetre, and other munitions of war were withheld from the Confederate armies. While the ports of the North were bustling with foreign trade, grass grew on the cobble-stoned streets along the waterfronts of Charleston and Savannah. Slow starvation aided the constant pounding of the Northern armies in reducing the South to subjection.

Had the Confederacy possessed but a few submarines of modern type this situation could not have persisted. Then, as to-day, neutral nations were eager to trade with both belligerents. There were then more neutrals whose interests would have compelled the observance of the laws of blockade, which in the present war are flagrantly violated by all belligerents with impunity. A submarine raid which would have sunk or driven away the blockading fleet at the entrance to a single harbour would have resulted in opening that harbour to the unrestricted uses of neutral ships until the blockade could be re-established and formal notice given to all powers—a formality which in those days, prior to the existence of cables, would have entailed weeks, perhaps months, of delay.

How serious such an interruption to the blockade was then considered was shown by the trepidation of the Union naval authorities

VICTORY IN THE CLOUDS

over the first victories of the *Merrimac* prior to the providential arrival of the *Monitor* in Hampton Roads. It was then thought that the Confederate ram would go straight to Wilmington, Charleston, and Savannah, destroy or drive away the blockaders, and open the Confederacy to the trade of the world.

Even then men dreamed of submarines, as indeed they have since the days of the American Revolution. Of the slow development of that engine of war to its present effectiveness we shall speak more fully in later chapters. Enough now to say that had the Confederacy possessed boats of the *U-53* type the story of our Civil War might have had a different ending. The device which the Allies have adopted to-day of blockading a port or ports by posting their ships several hundred miles away would have found no toleration among neutrals none too friendly to the United States, and vastly stronger in proportion to the power of this nation than all the neutrals to-day are to the strength of the Allies.

From the beginning of the Great War in Europe the fleets of the Teutonic alliance were locked up in port by the superior floating forces of the Entente. Such sporadic dashes into the arena of conflict as the one made by the German High Fleet, bringing on the Battle of Jutland, had but little bearing on the progress of the war. But the steady, persistent malignant activity of the German submarines had everything to do with it. They mitigated the rigidity of the British blockade by keeping the blockaders far from the ports they sought to seal. They preyed on the British fleets by sinking dreadnoughts, battleships, and cruisers in nearly all of the belligerent seas. If the British navy justified its costly power by keeping the German fleet practically imprisoned in its fortified harbours, the German submarines no less won credit and glory by keeping even that overwhelming naval force restricted in its movements, ever on guard, ever in a certain sense on the defensive. And meanwhile these underwater craft so preyed upon British food ships that in the days of the greatest submarine activity England was reduced to husbanding her stores of food with almost as great thrift and by precisely the same methods as did Germany suffering from the British blockade.

Aircraft and submarines! Twin terrors of the world's greatest war! The development, though by no means the final development, of dreams that men of many nations have dreamed throughout the centuries! They are two of the outstanding features of the war; two of its legacies to mankind. How much the legacy may be worth in peaceful times is yet to be determined. The aeroplane and the dirigible at any rate seem already to promise useful service to peaceful man. Already the flier is almost as common a spectacle in certain sections of our country as the automobile was fifteen years ago. The submarine, for economic reasons, promises less for the future in the way of peaceful service, notwithstanding the exploits of the *Deutschland* in the ocean-carrying trade. But perhaps it too will find its place in industry when awakened man shall be willing to spend as much treasure, as much genius, as much intelligent effort, and as much heroic self-sacrifice in organizing for the social good as in the last four years he has expended in its destruction.

CHAPTER 2

The Earliest Flying Men

The conquest of the air has been the dream of mankind for uncounted centuries. As far back as we have historic records we find stories of the attempts of men to fly. The earliest Greek mythology is full of aeronautical legends, and the disaster which befell Icarus and his wings of wax when exposed to the glare of the midsummer sun in Greece, is part of the schoolboy's task in Ovid. We find like traditions in the legendary lore of the Peruvians, the East Indians, the Babylonians, even the savage races of darkest Africa. In the Hebrew scriptures the chief badge of sanctity conferred on God's angels was wings, and the ability to fly. If we come down to the mythology of more recent times we find our pious ancestors in New England thoroughly convinced that the witches they flogged and hanged were perfectly able to navigate the air on a broomstick—thus antedating the Wrights' experiments with heavier-than-air machines by more than 250 years.

It is an interesting fact, stimulating to philosophical reflection, that in the last decade more has been done toward the conquest of the air, than in the twenty centuries preceding it, though during all that period men had been dreaming, planning, and experimenting upon contrivances for flight. Moreover when success came—or such measure of success as has been won—it came by the application of an entirely novel principle hardly dreamed of before the nineteenth century.

Some of the earlier efforts to master gravity and navigate the air are worthy of brief mention if only to show how persistent were the efforts from the earliest historic ages to accomplish this end. Passing over the legends of the time of mythology we find that many-sided genius, Leonardo da Vinci, early in the sixteenth century, not content with being a painter, architect, sculptor, engineer and designer of forts, offering drawings and specifications of wings which, fitted to men, he thought would enable them to fly. The sketches are still preserved in a museum at Paris. He modelled his wings on those of a bat and worked them with ropes passing over pulleys, the aviator lying prone, face downward, and kicking with both arms and legs with the vigour of a frog. There is, unhappily, no record that the proposition ever advanced beyond the literary stage—certainly none that Da Vinci himself thus risked his life. History records no one who kicked his way aloft with the Da Vinci device. But the manuscript which the projector left shows that he recognized the modern aviator's maxim, "There's safety in altitude." He says, in somewhat confused diction:

> The bird should with the aid of the wind raise itself to a great height, and this will be its safety; because although the revolutions mentioned may happen there is time for it to recover its equilibrium, provided its various parts are capable of strong resistance so that they may safely withstand the fury and impetus of the descent.

The fallacy that a man could, by the rapid flapping of wings of any sort, overcome the force of gravity persisted up to a very recent day, despite the complete mathematical demonstration by von Helmholtz in 1878 that man could not possibly by his own muscular exertions raise his own weight into the air and keep it suspended. Time after time the "flapping wings" were resorted to by ambitious aviators with results akin to those attained by Darius Green. One of the earliest was a French locksmith named Besnier, who had four collapsible planes on two rods balanced across his shoulders. These he vigorously moved up and down with his hands and feet, the planes opening like covers of a book as they came down, and closing as they came up. Besnier made no attempt to raise himself from the ground, but believed that once launched in the air from an elevation he could maintain himself, and glide gradually to earth at a considerable distance. It is said that he and one or two of his students did in a way accomplish this. Others, however, experimenting with the same method came to sorry disaster. Among these was an Italian friar whom King James IV. of Scotland had made Prior of Tongland. Equipped with a pair of large feather wings operated on the Besnier principle, he launched himself from the battlements of Stirling Castle in the presence of King James and his court. But gravity was too much for his apparatus, and turning over and over in mid-air he finally landed ingloriously on a manure heap—at that period of nascent culture a very common feature of the

The Fall of the Boche.

pleasure grounds of a palace. He had a soul above his fate however, for he ascribed his fall not to vulgar mechanical causes, but wholly to the fact that he had overlooked the proper dignity of flight by pluming his wings with the feathers of common barn-yard fowl instead of with plumes plucked from the wings of eagles!

In sharp competition with the aspiring souls who sought to fly with wings—the forerunners of the aeroplane devotees of to-day—were those who tried to find some direct lifting device for a car which should contain the aviators. Some of their ideas were curiously logical and at the same time comic. There was, for example, a priest, Le Père Galien of Avignon. He observed that the rarefied air at the summit of the Alps was vastly lighter than that in the valleys below. What then was to hinder carrying up empty sacks of cotton or oiled silk to the mountain tops, opening them to the lighter air of the upper ranges, and sealing them hermetically when filled by it. When brought down into the valleys they would have lifting power enough to carry tons up to the summits again. The good Father's education in physics was not sufficiently advanced to warn him that the effort to drag the balloons down into the valley would exact precisely the force they would exert in lifting any load out of the valley—if indeed they possessed any lifting power whatsoever, which is exceedingly doubtful.

Another project, which sounded logical enough, was based on the irrefutable truth that as air has some weight—to be exact 14.70 pounds for a column one inch square and the height of the earth's atmosphere—a vacuum must be lighter, as it contains nothing, not even air. Accordingly in the seventeenth century, one Francisco Lana, another priest, proposed to build an airship supported by four globes of copper, very thin and light, from which all the air had been pumped. The globes were to be twenty feet in diameter, and were estimated to have a lifting force of 2650 pounds. The weight of the copper shells was put at 1030 pounds, leaving a margin of possible weight for the car and its contents of 1620 pounds. It seemed at first glance a perfectly reasonable and logical plan. Unhappily one factor in the problem had been ignored. The atmospheric pressure on each of the globes would be about 1800 tons. Something more than a thin copper shell would be needed to resist this crushing force and an adequate increase in the strength of the shells would so enhance their weight as to destroy their lifting power.

To tell at length the stories of attempt and failure of the earliest dabblers in aeronautics would be unprofitable and uninteresting. Not

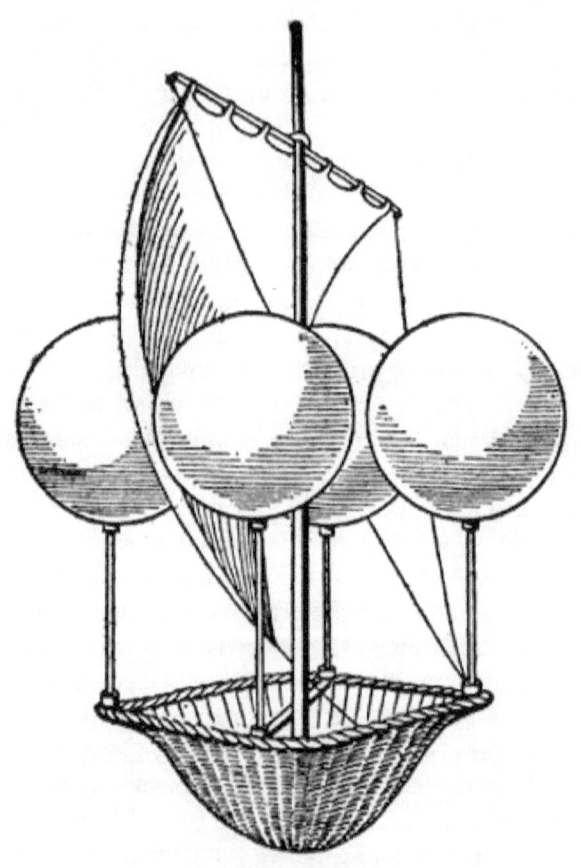

LANA'S VACUUM BALLOON

until the eighteenth century did the experimenters with lighter-than-air devices show any practical results. Not until the twentieth century did the advocates of the heavier-than-air machines show the value of their fundamental idea. The former had to discover a gaseous substance actually lighter, and much lighter, than the surrounding atmosphere before they could make headway. The latter were compelled to abandon wholly the effort to imitate the flapping of a bird's wings, and study rather the method by which the bird adjusts the surface of its wings to the wind and soars without apparent effort, before they could show the world any promising results.

Nearly every step forward in applied science is accomplished because of the observation by some thoughtful mind of some common phenomenon of nature, and the later application of those observations to some useful purpose.

It seems a far cry from an ancient Greek philosopher reposing peacefully in his bath to a modern Zeppelin, but the connection is direct. Every schoolboy knows the story of the sudden dash of Archimedes, stark and dripping from his tub, with the triumphant cry of "Eureka!"—"I have found it!" What he had found was the rule which governed the partial flotation of his body in water. Most of us observe it, but the philosophical mind alone inquired "Why?" Archimedes' answer was this rule which has become a fundamental of physics: "A body plunged into a fluid is subjected by this fluid to a pressure from below to above equal to the weight of the fluid displaced by the body." A balloon is plunged in the air—a fluid. If it is filled with air there is no upward pressure from below, but if it is filled with a gas lighter than air there is a pressure upward equal to the difference between the weight of that gas and that of an equal quantity of air. Upon that fact rests the whole theory and practice of ballooning.

The illustration of James Watt watching the steam rattle the cover of a teapot and from it getting the rudimentary idea of the steam engine is another case in point. Sometimes however the application of the hints of nature to the needs of man is rather ludicrously indirect. Charles Lamb gravely averred that because an early Chinaman discovered that the flesh of a pet pig, accidentally roasted in the destruction by fire of his owner's house, proved delicious to the palate, the Chinese for years made a practice of burning down their houses to get roast pig with "crackling." Early experimenters in aviation observed that birds flapped their wings and flew. Accordingly they believed that man to fly must have wings and flap them likewise. Not for hundreds of years did they observe that most birds flapped their wings only to get headway, or altitude, thereafter soaring to great heights and distances merely by adjusting the angle of their wings to the various currents of air they encountered.

In a similar way the earliest experimenters with balloons observed that smoke always ascended. "Let us fill a light envelope with smoke," said they, "and it will rise into the air bearing a burden with it." All of which was true enough, and some of the first balloonists cast upon their fires substances like sulphur and pitch in order to produce a thicker smoke, which they believed had greater lifting power than ordinary hot air.

In the race for actual accomplishment the balloonists, the advocates of lighter-than-air machines, took the lead at first. It is customary and reasonable to discard as fanciful the various devices and theories put forward by the experimenters in the Middle Ages and fix the beginning of practical aeronautical devices with the invention of hot-air balloons by the Montgolfiers, of Paris, in 1783.

The Montgolfier brothers, Joseph and Jacques, were paper-makers of Paris. The family had long been famous for its development of the paper trade, and the many ingenious uses to which they put its staple. Just as the tanners of the fabled town in the Middle Ages thought there was "nothing like leather" with which to build its walls and gates, thereby giving a useful phrase to literature, so the Montgolfiers thought of everything in terms of paper. Sitting by their big open fireplace one night, so runs the story, they noticed the smoke rushing up the chimney. "Why not fill a big paper bag with smoke and make it lift objects into the air?" cried one. The experiment was tried next day with a small bag and proved a complete success. A neighbouring housewife looked in, and saw the bag bumping about the ceiling, but rapidly losing its buoyancy as the smoke escaped.

"Why not fasten a pan below the mouth of the bag," said she, "and put your fire in that? Its weight will keep the bag upright, and when it rises will carry the smoke and the pan up with it."

Acting upon the hint the brothers fixed up a small bag which sailed up into the air beyond recapture. After various experiments a bag of mixed paper and linen thirty-five feet in diameter was inflated and released. It soared to a height of six thousand feet, and drifted before the wind a mile or more before descending. The ascent took place at Avonay, the home at the time of the Montgolfiers, and as every sort of publicity was given in advance, a huge assemblage including many officials of high estate gathered to witness it. A roaring fire was built in a pit over the mouth of which eight men held the great sack, which rolled, and beat about before the wind as it filled and took the form of a huge ball. The crowd was unbelieving and cynical, inclined to scoff at the idea that mere smoke would carry so huge a construction up into the sky. But when the signal was given to cast off, the balloon rose with a swiftness and majesty that at first struck the crowd dumb, then moved it to cheers of amazement and admiration. It went up six thousand feet and the Montgolfiers were at once elevated to almost an equal height of fame. The crowd which watched the experiment was wild with enthusiasm; the Montgolfiers elated with the first considerable victory over the force of gravity. They had demonstrated a principle and made their names immortal. What remained was to develop that principle and apply it to practical ends. That develop-

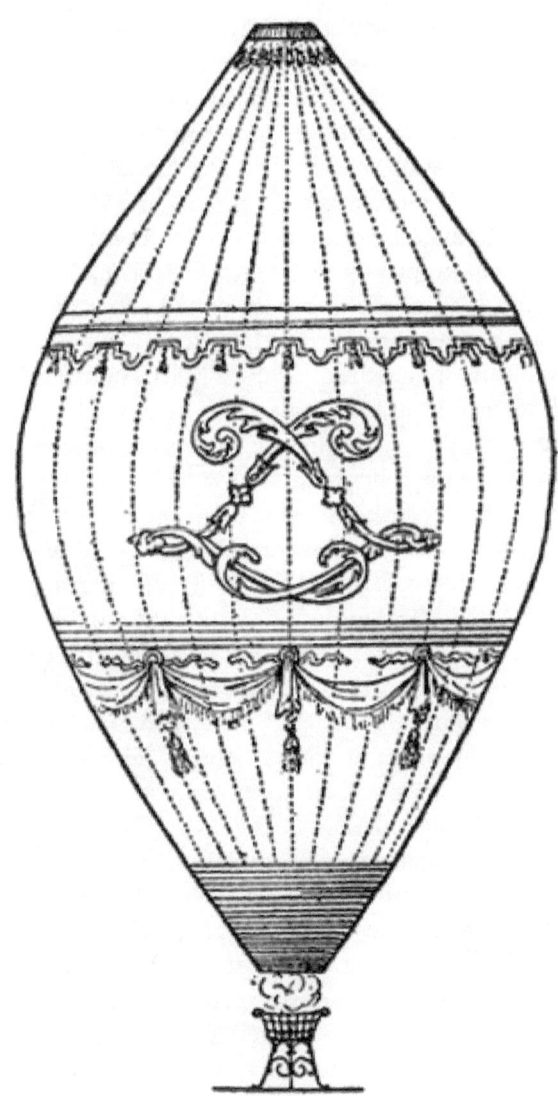

MONTGOLFIER'S EXPERIMENTAL BALLOON

ment, however, proceeded for something more than a century before anything like a practical airship was constructed.

But for the moment the attack on the forces which had kept the air virgin territory to man was not allowed to lag. In Paris public subscriptions were opened to defray the cost of a new and greater balloon. By this time it was known that hydrogen gas, or "inflammable air" as it was then called, was lighter than air. But its manufacture was then expensive and public aid was needed for the new experiment which would call at the outset for a thousand pounds of iron filings and 498 pounds of sulphuric acid wherewith to manufacture the gas.

The first experiment had been made in the provinces. This one was set for Paris, and in an era when the French capital was intellectually more alert, more eager for novelty, more interested in the advancement of physical science and in new inventions than ever in its long history of hospitality to the new idea. They began to fill the bag August 23, 1783 in the Place des Victoires, but the populace so thronged that square that two days later it was moved half filled to Paris's most historic point, the Champ de Mars. The transfer was made at midnight through the narrow dark streets of medieval Paris. Eye-witnesses have left descriptions of the scene. Torch-bearers lighted on its way the cortège the central feature of which was the great bag, half filled with gas, flabby, shapeless, monstrous, mysterious, borne along by men clutching at its formless bulk. The state had recognized the importance of the new device and *cuirassiers* in glittering breastplates on horseback, and halbardiers in buff leather on foot guarded it in its transit through the sleeping city. But Paris was not all asleep. An escort of the sensation-loving rabble kept pace with the guards. The cries of the quarters rose above the tramp of the armed men. Observers have recorded that the passing cab drivers were so affected by wonder that they clambered down from their boxes and with doffed hats knelt in the highway while the procession passed.

The ascension, which occurred two days later, was another moving spectacle. In the centre of the great square which has seen so many historic pageants, rose the swaying, quivering balloon, now filled to its full capacity of twenty-two thousand feet. Whether from the art instinct indigenous to the French, or some superstitious idea like that which impels the Chinese to paint eyes on their junks, the balloon was lavishly decorated in water colours, with views of rising suns, whirling planets, and other solar bodies amongst which it was expected to mingle.

Ranks of soldiers kept the populace at a distance, while within the sacred precincts strolled the king and the ladies and cavaliers of his court treading all unconsciously on the brink of that red terror soon to engulf the monarchy. The gas in the reeling bag was no more inflammable than the air of Paris in those days just before the Revolution. With a salvo of cannon the guy-ropes were released and the balloon vanished in the clouds.

Benjamin Franklin, at the moment representing in France the

A Rescue at Sea

American colonies then struggling for liberty, witnessed this ascension! "Of what use is a new-born child?" he remarked sententiously as the balloon vanished. 'Twas a saying worthy of a cautious philosopher. Had Franklin been in Paris in 1914 he would have found the child, grown to lusty manhood, a strong factor in the city's defence. It is worth noting by the way that so alert was the American mind at that period that when the news of the Montgolfiers' achievement reached Philadelphia it found David Rittenhouse and other members of the Philosophical Society already experimenting with balloons.

A curious sequel attended the descent of the Montgolfier craft which took place in a field fifteen miles from Paris. Long before the days of newspapers, the peasants had never heard of balloons, and this mysterious object, dropping from high heaven into their peaceful carrot patch affrighted them. Some fled. Others approached timidly, armed with the normal bucolic weapons—scythes and pitchforks. Attacked with these the fainting monster, which many took for a dragon, responded with loud hisses and emitted a gas of unfamiliar but most pestiferous odour. It suggested brimstone, which to the devout in turn implied the presence of Satan. With guns, flails, and all obtainable weapons they fell upon the emissary of the Evil One, beat him to the ground, crushed out of him the vile-smelling breath of his nostrils, and finally hitched horses to him and dragged him about the fields until torn to tatters and shreds.

When the public-spirited M. Charles who had contributed largely to the cost of this experiment came in a day or two to seek his balloon he found nothing but some shreds of cloth, and some lively legends of the prowess of the peasants in demolishing the devil's own dragon.

The government, far-sightedly, recognizing that there would be more balloons and useful ones, thereupon issued this proclamation for the discouragement of such bucolic valour:

> A discovery has been made which the government deems it wise to make known so that alarm may not be occasioned to the people. On calculating the different weights of inflammable and common air it has been found that a balloon filled with inflammable air will rise toward heaven until it is in equilibrium with the surrounding air; which may not happen till it has attained to a great height. Anyone who should see such a globe, resembling the moon in an eclipse, should be aware that far from being an alarming phenomenon it is only a machine

made of taffetas, or light canvas covered with paper, that cannot possibly cause any harm and which will some day prove serviceable to the wants of society.

Came now the next great step in the progress of aeronautics. It had been demonstrated that balloons could lift themselves. They had even been made to lift dumb animals and restore them to earth unhurt. But if the conquest of the air was to amount to anything, men must go aloft in these new machines. Lives must be risked to demonstrate a theory, or to justify a calculation. Aeronautics is no science for laboratory or library prosecution. Its battles must be fought in the sky, and its devotees must be willing to offer their lives to the cause. In that respect the science of aviation has been different from almost any subject of inquiry that has ever engaged the restless intellect of man, unless perhaps submarine navigation, or the invention of explosives. It cannot be prosecuted except with a perfect willingness to risk life. No doubt this is one of the reasons why practical results seemed so long in the coming. Nor have men been niggardly in this enforced sacrifice. Though no records of assured accuracy are available, the names of forty-eight aeronauts who gave up their lives in the century following the Montgolfiers' invention are recorded. That record ended in 1890. How many have since perished, particularly on the battlefields of Europe where aircraft are as commonplace as cannon, it is too early yet to estimate.

After the success of the ascension from the Champ de Mars, the demand at once arose for an ascension by a human being. It was a case of calling for volunteers. The experiments already made showed clearly enough that the balloon would rise high in air. Who would risk his life soaring one thousand feet or more above the earth, in a flimsy bag, filled with hot air, or inflammable gas, without means of directing its course or bringing it with certainty and safety back to a landing place? It was a hard question, and it is interesting to note that it was answered not by a soldier or sailor, not by an adventurer, or devil-may-care spirit, but by a grave and learned professor of physical science, Pilatre de Rozier. Presently he was joined in his enterprise by a young man of the fashionable world and sporting tastes, the Marquis d'Arlandes. Aristocratic Paris took up aviation in the last days of the eighteenth century, precisely as the American leisure class is taking it up in the first days of the twentieth.

The balloon for this adventure was bigger than its predecessors and for the first time a departure was taken from the spherical variety—the

MONTGOLFIER'S PASSENGER BALLOON

gas bag being seventy-four feet high, and forty-eight feet in diameter. Like the first Montgolfier balloons it was to be inflated with hot air, and the car was well packed with bundles of fuel with which the two aeronauts were to fill the iron brazier when its fires went down. The instinct for art and decoration, so strong in the French mind, had been given full play by the constructors of this balloon and it was painted with something of the gorgeousness of a circus poster.

A tremendous crowd packed the park near Paris whence the ascent was made. Always the spectacle of human lives in danger has a morbid attraction for curiosity seekers, and we have seen in our own days throngs attracted to aviation congresses quite as much in the expectation of witnessing some fatal disaster, as to observe the progress made in man's latest conquest over nature. But in this instance the occasion justified the widest interest. It was an historic moment—more epoch-making than those who gathered in that field in the environs of Paris could have possibly imagined. For in the clumsy, gaudy bag, rolling and tossing above a smoky fire lay the fundamentals of those great airships that, perfected by the persistence of Count Zeppelin, have crossed angry seas, breasted fierce winds, defied alike the blackest nights and the thickest fogs to rain their messages of death on the capital of a foe.

Contemporary accounts of this first ascension are but few, and those that have survived have come down to us in but fragmentary form. It was thought needful for two to make the ascent, for the car, or basket, which held the fire hung below the open mouth of the bag, and the weight of a man on one side would disturb the perfect equilibrium which it was believed would be essential to a successful flight. The Marquis d'Arlandes in a published account of the brief flight, which sounds rather as if the two explorers of an unknown element were not free from nervousness, writes:

> Our departure was at fifty-four minutes past one, and occasioned little stir among the spectators. Thinking they might be frightened and stand in need of encouragement I waved my arm.

This solicitude for the fears of the spectators, standing safely on solid earth while the first aeronauts sailed skywards, is characteristically Gallic. The marquis continues:

> M. de Rozier cried: "You are doing nothing, and we are not rising." I stirred the fire and then began to scan the river, but Pilatre again cried: "See the river. We are dropping into it!" We again urged the fire, but still clung to the river bed. Presently I heard a noise in the upper part of the balloon, which gave a shock as though it had burst. I called to my companion: "Are you dancing?" The balloon by this time had many holes burnt in it and using my sponge I cried that we must descend. My companion however explained that we were over Paris and must now cross it; therefore raising the fire once more we turned south till we passed the Luxembourg, when, extinguishing the flames, the balloon came down spent and empty.

If poor Pilatre played the part of a rather nervous man in this narrative he had the nerve still to go on with his aeronautical experiments to the point of death. In 1785 he essayed the crossing of the English Channel in a balloon of his own design, in which he sought to combine the principles of the gas and hot-air balloons. It appears to have been something like an effort to combine nitro-glycerine with an electric spark. At any rate the dense crowds that thronged the coast near Boulogne to see the start of the *Charles-Montgolfier*—as the balloon was named after the originators of the rival systems—saw it, after half an hour's drift out to sea, suddenly explode in a burst of flame. De Rozier and a friend who accompanied him were killed. A monument still recalls their fate, which however is more picturesquely recorded in the signs of sundry inns and cafés of the neighbourhood which offer refreshment in the name of *Les Aviateurs Perdus*.

Thereafter experimenters with balloons multiplied amazingly. The world thought the solution of the problem of flight had been found in the gas bag. Within two months a balloon capable of lifting eighteen tons and carrying seven passengers ascended three thousand feet at Lyons, and, though sustaining a huge rent in the envelope, because of the expansion of the gas at that height, returned to earth in safety. The fever ran from France to England and in 1784, only a year after the first Montgolfier experiments, Lunardi, an Italian aeronaut made an ascension from London which was viewed by King George III. and his ministers, among them William Pitt. But the early enthusiasm for ballooning quickly died down to mere curiosity. It became apparent to all that merely to rise into the air, there to be the helpless plaything of the wind, was but a useless and futile accomplishment. Pleasure seekers and mountebanks used balloons for their own purposes, but serious experimenters at once saw that if the invention of the balloon was to be of the slightest practical value some method must be devised for controlling and directing its flight. To this end some of the brightest intellects of the world directed their efforts, but it is hardly overstating the case to say that more than a century passed without any considerable progress toward the development of a dirigible balloon.

But even at the earlier time it was evident enough that the Quaker philosopher, from the American Colonies, not yet the United States, whose shrewd and inquiring disposition made him intellectually one of the foremost figures of his day, foresaw clearly the great possibilities of this new invention. In letters to Sir Joseph Banks, then President

CHARLES'S BALLOON

of the Royal Society of London, Franklin gave a lively account of the first three ascensions, together with some comments, at once suggestive and humorous, which are worth quoting:

> Some think Progressive Motion on the Earth may be advanc'd by it, and that a Running Footman or a Horse slung and suspended under such a Globe so as to have no more of Weight pressing the Earth with their Feet than Perhaps 8 or 10 Pounds, might with a fair Wind run in a straight Line across Countries as fast as that Wind, and over Hedges, Ditches and even Waters. It has been even fancied that in time People will keep

such Globes anchored in the Air to which by Pullies they may draw up Game to be preserved in the Cool and Water to be frozen when Ice is wanted. And that to get Money it will be contriv'd, by running them up in an Elbow Chair a Mile high for a guinea, etc., etc.

With his New England lineage Franklin could hardly have failed of this comparison:

A few Months since the Idea of Witches riding through the Air upon a broomstick, and that of Philosophers upon a Bag of Smoke would have appeared equally impossible and ridiculous.

To-day when aircraft are the eyes of the armies in the greatest war of history, and when it appears that, with the return of peace, the conquest of the air for the ordinary uses of man will be swiftly completed, Franklin's good-humoured plea for the fullest experimentation is worth recalling. And the touch of piety with which he concludes his argument is a delightful example of the whimsical fashion in which he often undertook to bolster up a mundane theory with a reference to things supernatural.

I am sorry this Experiment is totally neglected in England, where mechanic Genius is so strong. I wish I could see the same Emulation between the two Nations as I see between the two Parties here. Your Philosophy seems to be too bashful. In this Country we are not so much afraid of being laught at. If we do a foolish thing, we are the first to laugh at it ourselves, and are almost as much pleased with a *Bon Mot* or a *Chanson*, that ridicules well the Disappointment of a Project, as we might have been with its success. It does not seem to me a good reason to decline prosecuting a new Experiment which apparently increases the power of Man over Matter, till we can see to what Use that Power may be applied. When we have learnt to manage it, we may hope some time or other to find Uses for it, as men have done for Magnetism and Electricity, of which the first Experiments were mere Matters of Amusement.

This Experience is by no means a trifling one. It may be attended with important Consequences that no one can foresee. We should not suffer Pride to prevent our progress in Science.

Beings of a Rank and Nature far superior to ours have not disdained to amuse themselves with making and launching Bal-

A FRENCH OBSERVATION BALLOON ON FIRE

loons, otherwise we should never have enjoyed the Light of those glorious objects that rule our Day & Night, nor have had the Pleasure of riding round the Sun ourselves upon the Balloon we now inhabit.

B. *Franklin*

The earliest experimenters thought that oars might be employed to propel and direct a balloon. The immediate failure of all endeavours of this sort, led them, still pursuing the analogy between a balloon and a ship at sea, to try to navigate the air with sails. This again proved futile. It is impossible for a balloon, or airship to "tack" or manoeuvre in any way by sail power. It is in fact a monster sail itself, needing some other power than the wind to make headway or steerage way against the wind. The sail device was tested only to be abandoned. Only when a trail rope dragging along the ground or sea is employed does the sail offer sufficient resistance to the wind to sway the balloon's course this way or that. And a trailer is impracticable when navigating great heights.

For these reasons the development of the balloon lagged, until Count Zeppelin and M. Santos-Dumont consecrated their fortunes, their inventive minds, and their amazing courage to the task of perfecting a dirigible. In a book, necessarily packed with information concerning the rapid development of aircraft which began in the last decade of the nineteenth century and was enormously stimulated during the war of all the world, the long series of early experiments with balloons must be passed over hastily. Though interesting historically these experiments were futile. Beyond having discovered what could *not* be done with a balloon the practitioners of that form of aeronautics were little further along in 1898 when Count Zeppelin came along with the first plan for a rigid dirigible than they were when Blanchard in 1786, seizing a favourable gale drifted across the English Channel to the French shore, together with Dr. Jefferies, an American. It was just 124 years later that Bleriot, a Frenchman, made the crossing in an aeroplane independently of favouring winds. It had taken a century and a quarter to attain this independence.

In a vague way the earliest balloonists recognized that power, independent of wind, was necessary to give balloons steerage way and direction. Steam was in its infancy during the early days of ballooning, but the efforts to devise some sort of an engine light enough to be carried into the air were untiring. Within a year after the experi-

ROBERTS BROTHERS' DIRIGIBLE

ments of the Montgolfier brothers, the suggestion was made that the explosion of small quantities of gun-cotton and the expulsion of the resulting gases might be utilized in some fashion to operate propelling machinery. Though the suggestion was not developed to any useful point it was of interest as forecasting the fundamental idea of the gas engines of to-day which have made aviation possible—that is, the creation of power by a series of explosions within the motor.

In the effort to make balloons dirigible one of the first steps was to change the form from the spherical or pear-shaped bag to a cylindrical, or cigar-shape. This device was adopted by the brothers Robert in France as early as 1784. Their balloon further had a double skin or envelope, its purpose being partly to save the gas which percolated through the inner skin, partly to maintain the rigidity of the structure. As gas escapes from an ordinary balloon it becomes flabby, and can be driven through the air only with extreme difficulty. In the balloon of the Robert brothers air could from time to time be pumped into the space between the two skins, keeping the outer envelope always fully distended and rigid. In later years this idea has been modified by incorporating in the envelope one large or a number of smaller balloons or "balloonets," into which air may be pumped as needed.

The shape too has come to approximate that of a fish rather than a bird, in the case of balloons at least. "The head of a cod and the tail of a mackerel," was the way Marey-Monge, the French aeronaut described it. Though most apparent in dirigible balloons, this will be seen to be the favourite design for aeroplanes if the wings be stripped off, and the body and tail alone considered. Complete, these machines are not unlike a flying fish.

In England, Sir George Cayley, as early as 1810 studied and wrote largely on the subject of dirigibles but, though the English call him the "father of British aeronautics," his work seems to have been rather theoretical than practical. He did indeed demonstrate mathematically that no lifting power existed that would support the cumbrous steam-engine of that date, and tried to solve this dilemma by devising a gas engine, and an explosive engine. With one of the latter, driven by a series of explosions of gunpowder, each in a separate cell set off by a detonator, he equipped a flying machine which attained a sufficient height to frighten Cayley's coachman, whom he had persuaded to act as pilot. The rather unwilling aviator, fearing a loftier flight, jumped out and broke his leg. Though by virtue of this martyrdom his name should surely have descended to fame with that of Cayley it has been lost, together with all record of any later performances of the machine, which unquestionably embodied some of the basic principles of our modern aircraft, though it antedated the first of these by nearly a century.

We may pass over hastily some of the later experiments with dirigibles that failed. In 1834 the Count de Lennox built an airship 130 feet long to be driven by oars worked by man power. When the crowd that gathered to watch the ascent found that the machine was too heavy to ascend even without the men, they expressed their lively contempt for the inventor by tearing his clothes to tatters and smashing his luckless airship. In 1852, another Frenchman, Henry Giffard,

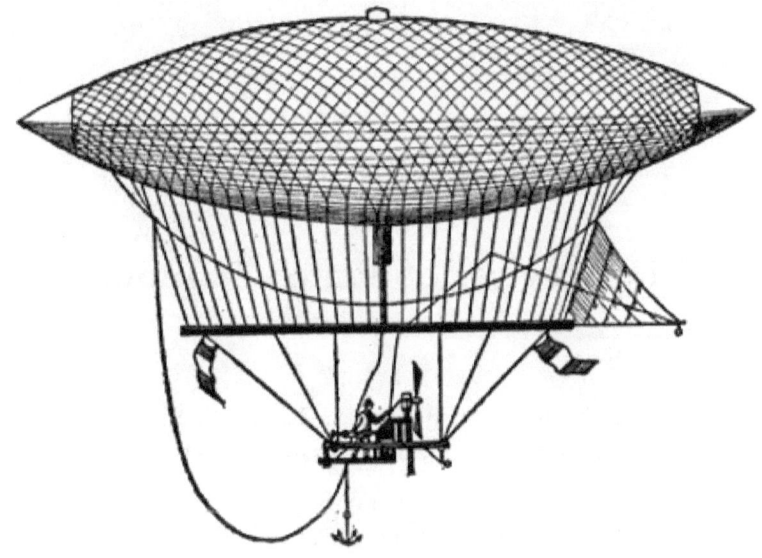

GIFFARD'S DIRIGIBLE.

built a cigar-shaped balloon 150 feet long by 40 feet in diameter, driven by steam. The engine weighed three hundred pounds and generated about 3 H.-P.—about one 200th as much power as a gas engine of equal weight would produce. Even with this slender power, however, Giffard attained a speed, independent of the wind, of from five to seven miles an hour—enough at least for steerage way. This was really the first practical demonstration of the possibilities of the mechanical propulsion of balloons. Several adaptations of the Giffard idea followed, and in 1883 Renard and Krebs, in a *fusiform* ship, driven by an electric motor, attained a speed of fifteen miles an hour. By this time inventive genius in all countries—save the United States which lagged in interest in dirigibles—was stimulated. Germany and France became the great protagonists in the struggle for precedence and in the struggle two figures stand out with commanding prominence—the Count von Zeppelin and Santos-Dumont, a young Brazilian resident in Paris who without official countenance consecrated his fortune to, and risked his life in, the service of aviation.

CHAPTER 3

The Services of Santos-Dumont

In his book *My Airships* the distinguished aviator A. Santos-Dumont tells of the ambition of his youth and its realization in later days:

> I cannot say at what age I made my first kites, but I remember how my comrades used to tease me at our game of "pigeon flies." All the children gather round a table and the leader calls out "Pigeon Flies! Hen flies! Crow flies! Bee flies!" and so on; and at each call we were supposed to raise our fingers. Sometimes, however, he would call out "Dog flies! Fox flies!" or some other like impossibility to catch us. If any one raised a finger then he was made to pay a forfeit. Now my playmates never failed to wink and smile mockingly at me when one of them called "Man flies!" for at the word I would always raise my finger very high, as a sign of absolute conviction, and I refused with energy to pay the forfeit. The more they laughed at me the happier I was, hoping that some day the laugh would be on my side.
>
> Among the thousands of letters which I received after winning the Deutsch prize (a prize offered in 1901 for sailing around the Eiffel Tower) there was one that gave me peculiar pleasure. I quote from it as a matter of curiosity:
>
>> Do you remember, my dear Alberto, when we played together 'Pigeon Flies!'? It came back to me suddenly when the news of your success reached Rio. 'Man flies!' old fellow! You were right to raise your finger, and you have just proved it by flying round the Eiffel Tower.
>>
>> They play the old game now more than ever at home; but the name has been changed, and the rules modified since October 19, 1901. They call it now 'Man flies!' and he who does not raise his finger at the word pays the forfeit.

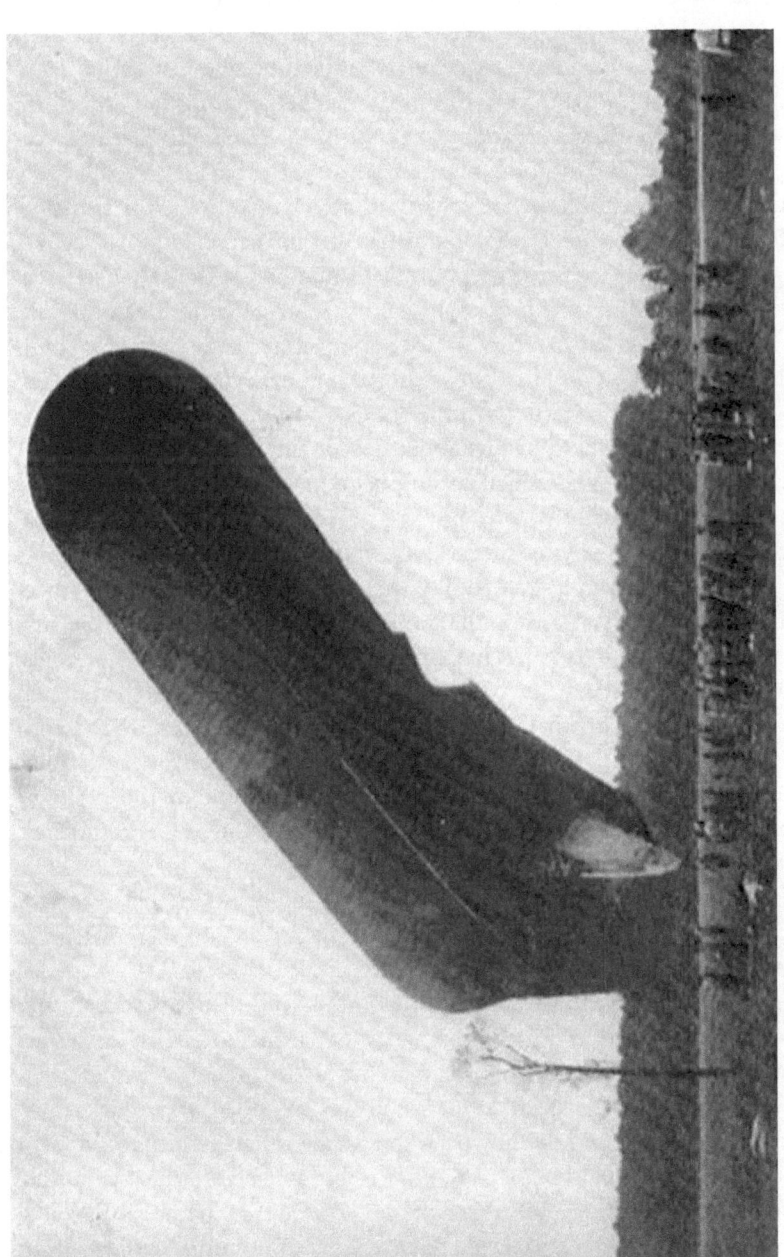

A British kite balloon: the open sack at the lower end catches the breeze and keeps the balloon steady

The story of Santos-Dumont affords a curious instance of a boy being obsessed by an idea which as a man he carried to its successful fruition. It offers also evidence of the service that may accrue to society from the devotion of a dilettante to what people may call a "fad," but what is in fact the germ of a great idea needing only an enthusiast with enthusiasm, brains, and money for its development. Because the efforts of Santos-Dumont always smacked of the amateur he has been denied his real place in the history of aeronautics, which is that of a fearless innovator, and a devoted worker in the cause.

Born on one of those great coffee plantations of Brazil, where all is done by machinery that possibly can be, Santos-Dumont early developed a passion for mechanics. In childhood he made toy aeroplanes. He confesses that his favourite author was Jules Verne, that literary idol of boyhood, who while writing books as wildly imaginative as any dime tale of redskins, or nickel novel of the doings of "Nick Carter" had none the less the spirit of prophecy that led him to forecast the submarine, the automobile, and the navigation of the air. At fifteen Santos-Dumont saw his first balloon and marked the day with red.

I too desired to go ballooning (he writes). In the long sun-bathed Brazilian afternoons, when the hum of insects, punctuated by the far-off cry of some bird lulled me, I would lie in the shade of the veranda and gaze into the fair sky of Brazil where the birds fly so high and soar with such ease on their great outstretched wings; where the clouds mount so gaily in the pure light of day, and you have only to raise your eyes to fall in love with space and freedom. So, musing on the exploration of the aerial ocean, I, too, devised airships and flying-machines in my imagination.

From dreaming, the boy's ambitions rapidly developed into actions. Good South Americans, whatever the practice of their northern neighbours, do not wait to die before going to Paris. At the age of eighteen the youth found himself in the capital of the world. To his amazement he found that the science of aeronautics, such as it was, had stopped with Giffard's work in 1852. No dirigible was to be heard of in all Paris. The antiquated gas ball was the only way to approach the upper air. When the boy tried to arrange for an ascension the balloonist he consulted put so unconscionable a price on one ascent that he bought an automobile instead—one of the first made, for this was in 1891—and with it returned to Brazil. It was not until six years later that, his ambition newly fired by reading of Andrée's plans for reaching the Pole in a balloon, Santos-Dumont took up anew his ambition

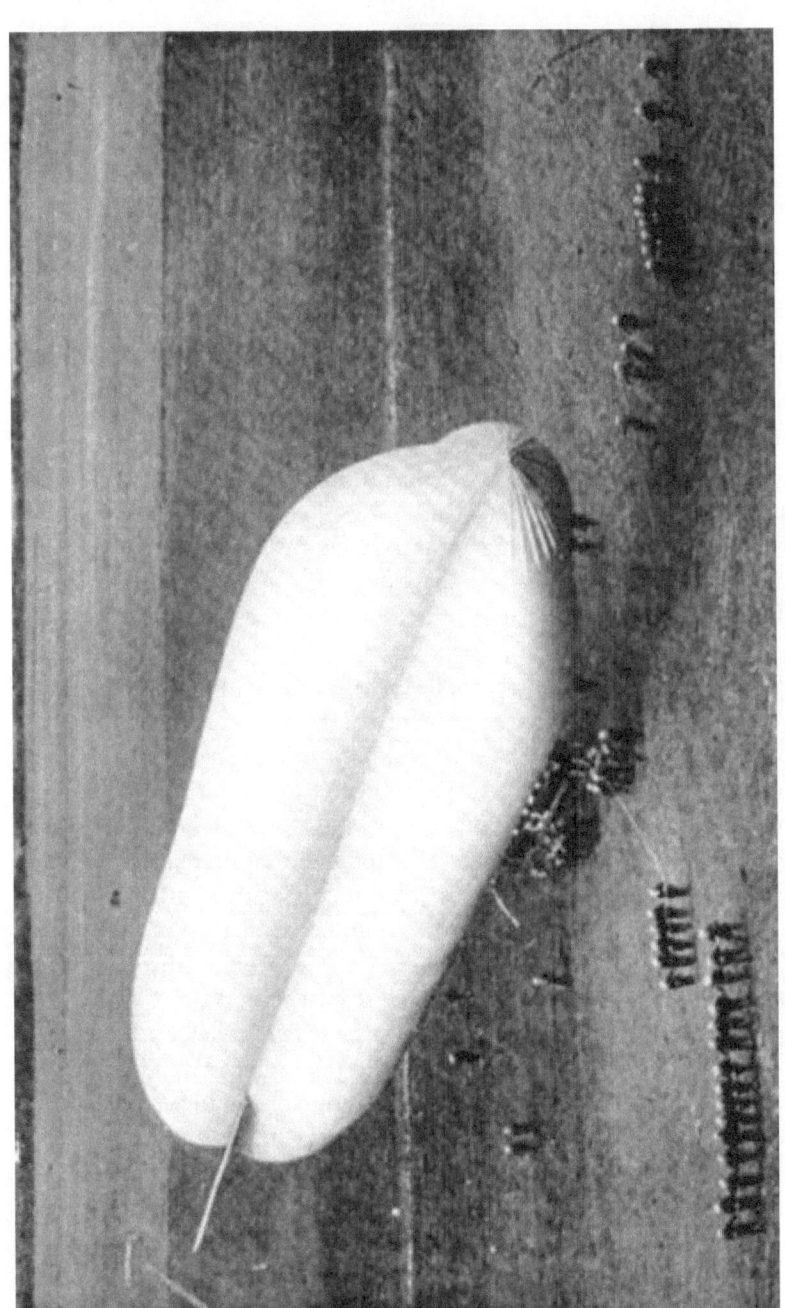

A British "Blimp" photographed from above

to become an aviator. His own account of his first ascent does not bear precisely the hall-mark of the enthusiast too rapt in ecstasy to think of common things. "I had brought up," he notes gravely, "a substantial lunch of hard-boiled eggs, cold roast beef and chicken, cheese, ice cream, fruits and cakes, champagne, coffee, and chartreuse!"

The balloon with its intrepid voyagers nevertheless returned to earth in safety.

A picturesque figure, an habitué of the clubs and an eager sportsman, Santos-Dumont at once won the liking of the French people, and attracted attention wherever people gave thought to aviation. Liberal in expenditure of money, and utterly fearless in exposing his life, he pushed his experiments for the development of a true dirigible tirelessly. Perhaps his major fault was that he learned but slowly from the experiences of others. He clung to the spherical balloon long after the impossibility of controlling it in the air was accepted as unavoidable by aeronauts. But in 1898 having become infatuated with the performances of a little sixty-six pound tricycle motor he determined to build a cigar-shaped airship to fit it, and with that determination won success.

Amateur he may have been, was indeed throughout the greater part of his career as an airman. Nevertheless Santos-Dumont has to his credit two very notable achievements.

He was the first constructor and pilot of a dirigible balloon that made a round trip, that is to say returned to its starting place after rounding a stake at some distance—in this instance the Eiffel Tower, 3½ miles from St. Cloud whence Santos-Dumont started and whither he returned within half an hour, the time prescribed.

This was not, indeed, the first occasion on which a round trip, necessitating operation against the wind on at least one course, had been made. In 1884 Captain Renard had accomplished this feat for the first time with the fish-shaped balloon *La France*, driven by an electric motor of nine horse-power. But though thus antedated in his exploit, Santos-Dumont did in fact accomplish more for the advancement and development of dirigible balloons. To begin with he was able to use a new and efficient form of motor destined to become popular, and capable, as the automobile manufacturers later showed, of almost illimitable development in the direction of power and lightness. Except for the gasoline engine, developed by the makers of motor cars, aviation to-day would be where it was a quarter of a century ago.

Moreover by his personal qualities, no less than by his successful demonstration of the possibilities inherent in the dirigible, Santos-Dumont persuaded the French Government to take up aeronautics again, after abandoning the subject as the mere fad of a number of visionaries.

Turning from balloons to aeroplanes the Brazilian was the first aviator to make a flight with a heavier-than-air machine before a body of judges. This triumph was mainly technical. The Wrights had made an equally notable flight almost a year before but not under conditions that made it a matter of scientific record.

But setting aside for the time the work done by Santos-Dumont with machines heavier than air, let us consider his triumphs with balloons at the opening of his career. He had come to France about forty years after Henry Giffard had demonstrated the practicability of navigating a balloon 144 feet long and 34 feet in diameter with a three-horse-power steam-engine. But no material success attended this demonstration, important as it was, and the inventor turned his attention to captive balloons, operating one at the Paris Exposition of 1878 that took up forty passengers at a time. There followed Captain Renard to whose achievement we have already referred. He had laid down as the fundamentals of a dirigible balloon these specifications:

A cigar, or fishlike shape.

An internal sack or ballonet into which air might be pumped to replace any lost gas, and maintain the shape of the balloon.

A keel, or other longitudinal brace, to maintain the longitudinal stability of the balloon and from which the car containing the motor might be hung.

A propeller driven by a motor, the size and power of both to be as great as permitted by the lifting power of the balloon.

A rudder capable of controlling the course of the ship.

Santos-Dumont adopted all of these specifications, but added to them certain improvements which gave his airships—he built five of them before taking his first prize—notable superiority over that of Renard. To begin with he had the inestimable advantage of having the gasoline motor. He further lightened his craft by having the envelope made of Japanese silk, in flat defiance of all the builders of balloons who assured him that the substance was too light and its use would be suicidal. "All right," said the innovator to his favourite constructor, who refused to build him a balloon of that material, "I'll build it myself." In the face of this threat the builder capitulated. The balloon was built, and the silk proved to be the best fabric available at that time for the purpose. A keel made of strips of pine banded together with aluminium wire formed the backbone of the Santos-Dumont craft, and from it depended the car about one quarter of the length of the balloon and hung squarely amidships. The idea of this keel occurred to the inventor while pleasuring at Nice. Later it saved his life.

One novel and exceedingly simple device bore witness to the ingenuity of the inventor. He had noticed in his days of free ballooning that to rise the aeronaut had to throw out sand-ballast; to descend he had to open the valves and let out gas. As his supply of both gas and sand was limited it was clear that the time of his flight was necessarily curtailed every time he ascended or descended. Santos-Dumont thought to husband his supplies of lifting force and of ballast, and make the motor raise and lower the ship. It was obvious that the craft would go whichever way the bow might be pointed, whether up or down. But how to shift the bow? The solution seems so simple that one wonders it ever perplexed aviators. From the peak of the bow and stern of his craft Santos-Dumont hung long ropes caught in the centre by lighter ropes by which they could be dragged into the car. In the car was carried a heavy bag of sand, which so long as it was there held the ship in a horizontal plane. Was it needful to depress the bow? Then the bow rope was hauled in, the bag attached, and swung out to a position where it would pull the forward tip of the delicately adjusted gas bag toward the earth. If only a gentle inclination was desired the bag was not allowed to hang directly under the bow, but was held at a point somewhere between the car and the bow so that the pull would be diagonal and the great cylinder would be diverted but little from the horizontal. If it were desired to ascend, a like manipulation of the ballast on the stern rope would depress the stern and point the bow upwards. For slight changes in direction it was not necessary even to attach the sand bag. Merely drawing the rope into the car and thus changing the line of its "pull" was sufficient.

The Deutsch prize which stimulated Santos-Dumont to his greatest achievements with dirigibles was a purse of twenty thousand dollars, offered by Mr. Henry Deutsch, a wealthy patron of the art of aviation. Not himself an aviator, M. Deutsch greatly aided the progress of the air's conquest. Convinced that the true solution of the problem lay in development of the gasoline engine, he expended large sums in developing and perfecting it. When he believed it was sufficiently developed to solve the problem of directing the flight of balloons he offered his prize for the circuit of the Eiffel Tower. The conditions of the contest were not easy. The competitor had to sail from the Aero Club at St. Cloud, pass twice over the Seine which at that point makes an abrupt bend, sail over the Bois de Boulogne, circle the Tower, and return to the stopping place within a half an hour. The distance was about seven miles, and it is noteworthy that in his own comment on the test Santos-Dumont complains that that required an average speed of fifteen miles an hour of which he could not be sure with his balloon. To-day dirigibles make sixty miles an hour, and aeroplanes not infrequently reach 130 miles. Moreover there could be no picking of a day on which atmospheric conditions were especially good. Mr. Deutsch had stipulated that the test must be made in the presence of a Scientific Commission whose members must be notified twenty-four hours in advance. None could tell twenty-four hours ahead what the air might be like, and as for utilizing the aviator's most favourable hour, the calm of the dawn, M. Santos-Dumont remarked: "The duellist may call out his friends at that sacred hour, but not the airship captain."

The craft with which the Brazilian first strove to win the Deutsch prize he called *Santos-Dumont No. V.* It was a cylinder, sharp at both ends, 109 feet long and driven by a 12-horse-power motor. A new feature was the use of piano wire for the support of the car, thus greatly reducing the resistance of the air which in the case of the old cord suspensions was almost as great as that of the balloon itself. Another novel feature was water ballast tanks forward and aft on the balloon itself and holding together twelve gallons. By pulling steel wires in the car the aviator could open the stop-cocks. The layman scarcely appreciates the very slight shift in ballast which will affect the stability of a dirigible. The shifting of a rope a few feet from its normal position, the dropping of two handfuls of sand, or release of a cup of water will do it. A humorous writer describing a lunch with Santos-Dumont in the air says: "Nothing must be thrown overboard, be it a bottle, an empty box or a chicken bone without the pilot's permission."

After unofficial tests of his *No. 5* in one of which he circled the Tower without difficulty, Santos-Dumont summoned the Scientific Commission for a test. In ten minutes he had turned the Tower, and started back against a fierce head-wind, which made him ten minutes late in reaching the time-keepers. Just as he did so his engine failed, and after drifting for a time his ship perched in the top of a chestnut tree on the estate of M. Edmond Rothschild. Philosophical as ever the aeronaut clung to his craft, dispatched an excellent lunch which the Princess Isabel, Comtesse d'Eu, daughter of Dom Pedro, the deposed Emperor of Brazil, sent to his eyrie in the branches, and finally extricated himself and his balloon—neither much the worse for the accident. He had failed but his determination to win was only whetted.

The second trial for the Deutsch prize like the first ended in failure, but that failure was so much more dramatic even than the success which attended the third effort that it is worth telling and can best be told in M. Santos-Dumont's own words. The quotation is from his memoir, *My Airships*:

> And now I come to a terrible day—8th of August, 1901. At 6:30 A.M. in presence of the Scientific Commission of the Aero Club, I started again for the Eiffel Tower.
>
> I turned the tower at the end of nine minutes and took my way back to St. Cloud; but my balloon was losing hydrogen through one of its two automatic gas valves whose spring had been accidentally weakened.
>
> I had perceived the beginning of this loss of gas even before reaching the Eiffel Tower, and ordinarily, in such an event, I should have come at once to earth to examine the lesion. But here I was competing for a prize of great honour and my speed had been good. Therefore I risked going on.
>
> The balloon now shrunk visibly. By the time I had got back to the fortifications of Paris, near La Muette, it caused the suspension wires to sag so much that those nearest to the screw-propeller caught in it as it revolved.
>
> I saw the propeller cutting and tearing at the wires. I stopped the motor instantly. Then, as a consequence, the airship was at once driven back toward the tower by the wind which was strong.
>
> At the same time I was falling. The balloon had lost much gas. I might have thrown out ballast and greatly diminished the fall, but then the wind would have time to blow me back on the Eiffel Tower. I therefore preferred to let the airship go down as it was going. It may have seemed a terrific fall to those who watched it from the ground but to me the worst detail was the airship's lack of equilibrium. The half-empty balloon, fluttering its empty end as an elephant waves his trunk, caused the air-

A KITE BALLOON RISING FROM THE HOLD OF A SHIP

ship's stern to point upward at an alarming angle. What I most feared therefore was that the unequal strain on the suspension wires would break them one by one and so precipitate me to the ground.

Why was the balloon fluttering an empty end causing all this extra danger? How was it that the rotary ventilator was not fulfilling its purpose in feeding the interior air balloon and in this manner swelling out the gas balloon around it? The answer must be looked for in the nature of the accident. The rotary ventilator stopped working when the motor itself stopped, and I had been obliged to stop the motor to prevent the propeller from tearing the suspension wires near it when the balloon first began to sag from loss of gas. It is true that the ventilator which was working at that moment had not proved sufficient to prevent the first sagging. It may have been that the interior balloon refused to fill out properly. The day after the accident when my balloon constructor's man came to me for the plans of a *No. 6* balloon envelope I gathered from something he said that the interior balloon of *No. 5"* not having been given time for its varnish to dry before being adjusted, might have stuck together or stuck to the sides or bottom of the outer balloon. Such are the rewards of haste.

I was falling. At the same time the wind was carrying me toward the Eiffel Tower. It had already carried me so far that I was expecting to land on the Seine embankment beyond the Trocadero. My basket and the whole of the keel had already passed the Trocadero hotels, and had my balloon been a spherical one it would have cleared the building. But now at the last critical moment, the end of the long balloon that was still full of gas came slapping down on the roof just before clearing it. It exploded with a great noise; struck after being blown up. This was the terrific explosion described in the newspaper of the day.

I had made a mistake in my estimate of the wind's force, by a few yards. Instead of being carried on to fall on the Seine embankment, I now found myself hanging in my wicker basket high up in the courtyard of the Trocadero hotels, supported by my airship's keel, that stood braced at an angle of about forty-five degrees between the courtyard wall above and the roof of a lower construction farther down. The keel, in spite of my weight, that of the motor and machinery, and the shock it had received in falling, resisted wonderfully. The thin pine scantlings and piano wires of Nice (the town where the idea of a keel first suggested itself) had saved my life!

After what seemed tedious waiting, I saw a rope being lowered to me from the roof above. I held to it and was hauled up, when I perceived my rescuers to be the brave firemen of Paris. From their station at Passy they had been watching the flight of the airship. They had seen my fall and immediately hastened to the spot. Then, having rescued me, they proceeded to rescue the airship.

The operation was painful. The remains of the balloon envelope and the suspension wires hung lamentably; and it was impossible to disengage them except in strips and fragments!

The later balloon *No. VI.* with which Santos-Dumont won the Deutsch prize may fairly be taken as his conception of the finished type of dirigible for one man. In fact his aspirations never soared as high as those of Count Zeppelin, and the largest airship he ever planned—called the *Omnibus*—carried only four men. It is probable that the diversion of his interest from dirigibles to aeroplanes had most to do with his failure to carry his development further than he did. *No. VI.* was 108 feet long, and 20 feet in diameter with an eighteen-horsepower gasoline engine which could drive it at about nineteen miles an hour. Naturally the aeronaut's first thought in his new construction was of the valves. The memory of the anxious minutes spent perched on the window-sill of the Trocadero Hotel or dangling like a spider at the end of the firemen's rope were still fresh. The ballonet which had failed him in *No. V.* was perfected in its successor. Notwithstanding the care with which she was constructed the prize-winner turned out to be a rather unlucky ship. On her trial voyage she ran into a tree and was damaged, and even on the day of her greatest conquest she behaved badly. The test was made on October 1, 1901. The aeronaut had rounded the Tower finely and was making for home when the motor began to miss and threatened to stop altogether. While Santos-Dumont was tinkering with the engine, leaving the steering wheel to itself, the balloon drifted over the Bois de Boulogne. As usual the cool air from the wood caused the hydrogen in the balloon to contract and the craft dropped until it appeared the voyage would end in the tree tops. Hastily shifting his weights the aeronaut forced the prow of the ship upwards to a sharp angle with the earth. Just at this moment the reluctant engine started up again with such vigour that for a moment the ship threatened to assume a perpendicular position, pointing straight up in the sky. A cry went up from the spectators below who feared a dire catastrophe was about to end a voyage which promised success. But with incomparable *sang-froid* the young Brazilian manipulated the weights, restored the ship to the horizontal again without stopping the engines, and reached the finishing stake in time to win the prize. Soon after it was awarded him the Brazilian Government presented him with another substantial prize, together with a gold medal bearing the words: *Por ceos nunca d'antes navegados* ("Through heavens hitherto unsailed").

In a sense the reference to the heavens is a trifle over-rhetorical. Santos-Dumont differed from all aviators (or pilots of aeroplanes) and most navigators of dirigibles in always advocating the strategy of staying near the ground. In his flights he barely topped the roofs of the houses, and in his writings he repeatedly refers to the sense of safety that came to him when he knew he was close to the tree tops of a forest. This may have been due to the fact that in his very first flight in a dirigible he narrowly escaped a fatal accident due to flying too high. As he descended, the gas which had expanded now contracted. The balloon began to collapse in the middle. Cords subjected to unusual stress began to snap. The air pump, which should have pumped the ballonet full of air to keep the balloon rigid failed to work. Seeing that he was about to fall into a field in which his drag rope was already trailing the imperilled airman had a happy thought. Some boys were there flying kites. He shouted to them to seize his rope and run against the wind. The balloon responded to the new force like a kite. The rapidity of its fall was checked, and its pilot landed with only a serious shaking.

But thereafter Santos-Dumont preached the maxim—rare among airmen—"Keep near the ground. That way lies safety!" Most aviators however, prefer the heights of the atmosphere, as the sailor prefers the wide and open sea to a course near land.

After winning the Deutsch prize, Santos-Dumont continued for a time to amuse himself with dirigibles. I say "amuse" purposely, for never did serious aeronaut get so much fun out of a rather perilous pastime as he. In his *No. IX* he built the smallest dirigible ever known. The balloon had just power enough to raise her pilot and sixty-six pounds more beside a three-horse-power motor. But she attained a speed of twelve miles an hour, was readily handled, and it was her owner's dearest delight to use her for a taxicab, calling for lunch at the cafés in the Bois, and paying visits to friends upon whom he looked in, literally, at their second-story windows. He ran her in and out of her hangar as one would a motor-car from its garage. One day he sailed down the Avenue des Champs Élysées at the level of the second-and third-story windows of the palaces that line that stately street. Coming to his own house he descended, made fast, and went in to *déjeuner*, leaving his aerial cab without. In the city streets he steered mainly by aid of a guide rope trailing behind him. With this he turned sharp corners, went round the Arc de Triomphe, and said: "I might have guide-roped under it had I thought myself worthy." On occasion he picked up children in the streets and gave them a ride.

Though before losing his interest in dirigibles Santos-Dumont carried the number of his construction up to ten, he cannot be said to have devised any new and useful improvements after his "*No. VI.*" The largest of his ships was "*No. X.*," which had a capacity of eighty thousand cubic feet—about ten times the size of the little runabout with which he played pranks in Paris streets. In this balloon he placed partitions to prevent the gas shifting to one part of the envelope, and to guard against losing it all in the event of a tear. The same principle was fundamental in Count Zeppelin's airships. In 1904 he brought a dirigible to the United States expecting to compete for a prize at the St. Louis Exposition. But while suffering exasperating delay from the red-tape which enveloped the exposition authorities, he discovered one morning that his craft had been mutilated almost beyond repair in its storage place. In high dudgeon he left at once for Paris. The explanation of the malicious act has never been made clear, though many Americans had an uneasy feeling that the gallant and sportsman-like Brazilian had been badly treated in our land. On his return to Paris he at once began experimenting with heavier-than-air machines. Of his work with them we shall give some account later.

Despite his great personal popularity the airship built by Santos-Dumont never appealed to the French military authorities. Probably this was largely due to the fact that he never built one of a sufficient size to meet military tests. The amateur in him was unconquerable. While von Zeppelin's first ship was big enough to take the air in actual war the Frenchman went on building craft for one or two men—good models for others to seize and build upon, but nothing which a war office could actually adopt. But he served his country well by stimulating the creation of great companies who built largely upon the foundations he had laid.

First and greatest of these was the company formed by the Lebaudy Brothers, wealthy sugar manufacturers. Their model was semi-rigid, that is, provided with an inflexible keel or floor to the gas bag, which was cigar shaped. The most successful of the earlier ships was 190 feet long, with a car suspended by cables ten feet below the balloon and carrying the twin motors, together with passengers and supplies. Although it made many voyages without accident, it finally encountered what seems to be the chief peril of dirigible balloons, being torn from its moorings at Châlons and dashed against trees to the complete demolition of its envelope. Repaired in eleven weeks she was taken over by the French Department of War, and was in active service at the beginning of the war. Her two successors on the company's building ways were less fortunate. *La Patrie*, after many successful trips, and manoeuvres with the troops, was insecurely moored at Verdun, the famous fortress where she was to have been permanently stationed. Came up a heavy gale. Her anchors began to drag. The bugles sounded and the soldiers by hundreds rushed from the fort to aid. Hurled along by the wind she dragged the soldiers after her. Fearing disaster to the men the commandant reluctantly ordered them to let go. The ship leaped into the black upper air and disappeared. All across France, across that very country where in 1916 the trenches cut their ugly zigzags from the Channel to the Vosges, she drifted unseen. By morning she was flying over England and Wales. Ireland caught a glimpse of her and days thereafter sailors coming into port told of a curious yellow mass, seemingly flabby and disintegrating like the carcass of a whale, floating far out at sea.

Her partner ship *La République* had a like tragic end. She too made many successful trips, and proved her stability and worth. But one day while manoeuvring near Paris one of her propellers broke and tore a great rent in her envelope. As the *Titanic*, her hull ripped open by an iceberg, sunk with more than a thousand of her people, so this airship, wounded in a more unstable element, fell to the ground killing all on board.

Two airships were built in France for England in 1909. One, the *Clement-Bayard II.*, was of the rigid type and built for the government; the other, a *Lebaudy*, was non-rigid and paid for by popular subscriptions raised in England by the *Morning Post*. Both were safely delivered near London having made their voyages of approximately 242 miles each at a speed exceeding forty miles an hour. These were the first airships acquired for British use.

In the United States the only serious effort to develop the dirigible prior to the war, and to apply it to some definite purpose, was made not by the government but by an individual. Mr. Walter Wellman, a distinguished journalist, fired by the effort of Andrée to reach the North Pole in a drifting balloon, undertook a similar expedition with a dirigible in 1907. A balloon was built 184 feet in length and 52 feet in diameter, and was driven by a seventy-to eighty-horse-power motor. A curious feature of this craft was the guide rope or, as Wellman called it, the equilibrator, which was made of steel, jointed and hollow. At the lower end were four steel cylinders carrying wheels and so arranged that they would float on water or trundle along over the roughest ice. The idea was that the equilibrator would serve like a guide rope, trailing on the water or ice when the balloon hung low, and increasing the power of its drag if the balloon, rising higher, lifted a greater part of its length into the air. Wellman had every possible appliance to contribute to the safety of the airship, and many believe that had fortune favoured him the glory of the discovery of the Pole would have been his. Unhappily he encountered only ill luck. One season he spent at Dane's Island, near Spitzenberg whence Andrée had set sail, waiting vainly for favourable weather conditions. The following summer, just as he was about to start, a fierce storm destroyed his balloon shed and injured the balloon. Before necessary repairs could be accomplished Admiral Peary discovered the Pole and the purpose of the expedition was at an end. Wellman, however, had become deeply interested in aeronautics and, balked in one ambition, set out to accomplish another. With the same balloon somewhat remodelled he tried to cross the Atlantic, setting sail from Atlantic City, N. J., October 16, 1911. But the device on which the aeronaut most prided himself proved his undoing. The equilibrator, relied upon both for storage room and as a regulator of the altitude of the ship, proved a fatal attachment. In even moderate weather it bumped over the waves and racked the structure of the balloon with its savage tugging until the machinery broke down and the adventurers were at the mercy of the elements. Luckily for them after they had been adrift for seventy-two hours, and travelled several hundred miles they were rescued by the British steamer *Trent*. Not long after Wellman's chief engineer Vanniman sought to cross the Atlantic in a similar craft but from some unexplained cause she blew up in mid-air and all aboard were lost.

Neither Great Britain nor the United States has reason to be proud of the attitude of its government towards the inventors who were struggling to subdue the air to the uses of man. Nor has either reason to boast much of its action in utterly ignoring up to the very day war broke that aid to military service of which Lord Kitchener said, "One aviator is worth a corps of cavalry." It will be noted that to get its first effective dirigible Great Britain had to rely upon popular subscriptions drummed up by a newspaper. That was in 1909. To-day, in 1917, the United States has only one dirigible of a type to be considered effective in the light of modern standards, though our entrance upon the war has caused the beginning of a considerable fleet. In aviation no less than in aerostatics the record of the United States is negligible. Our country did indeed produce the Wright Brothers, pioneers and true conquerors of the air with aeroplanes. But even they were forced to go to France for support and indeed for respectful attention.

So far as the development of dirigible balloons is concerned there is no more need to devote space to what was done in England and the United States than there was for the famous chapter on Snakes in Iceland.

CHAPTER 4

The Count von Zeppelin

The year that witnessed the first triumphs of Santos-Dumont saw also the beginning of the success of his great German rival, the Count von Zeppelin. These two daring spirits, struggling to attain the same end, were alike in their enthusiasm, their pertinacity, and their devotion to the same cause. Both were animated by the highest patriotism. Santos-Dumont offered his fleet to France to be used against any nation except those of the two Americas. He said: "It is in France that I have met with all my encouragement; in France and with French material I have made all my experiments. I excepted the two Americas because I am an American."

Count Zeppelin for his part, when bowed down in apparent defeat and crushed beneath the burden of virtual bankruptcy, steadily refused to deal with agents of other nations than Germany—which at that time was turning upon him the cold shoulder. He declared that his genius had been exerted for his own country alone, and that his invention should be kept a secret from all but German authorities. A secret it would be to-day, except that accident and the fortunes of war revealed the intricacies of the Zeppelin construction to both France and England.

Santos-Dumont had the fire, enthusiasm, and resiliency of youth; Zeppelin, upon whom age had begun to press when first he took up aeronautics, had the dogged pertinacity of the Teuton. Both were rich at the outset, but Zeppelin's capital melted away under the demands of his experimental workshops, while the ancestral coffee lands of the Brazilian never failed him.

Of the two Zeppelin had the more obstinacy, for he held to his plan of a rigid dirigible balloon even in face of its virtual failure in the supreme test of war. Santos-Dumont was the more alert intellectually for he was still in the flood tide of successful demonstration with his balloons when he saw and grasped the promise of the aeroplane and shifted his activities to that new field in which he won new laurels.

Zeppelin won perhaps the wider measure of immediate fame, but whether enduring or not is yet to be determined. His airships impressive, even majestic as they are, have failed to prove their worth in war, and are yet to be fully tested in peace. That they remain a unique type, one which no other individual nor any other nation has sought to copy, cannot be attributed wholly to the jealousy of possible rivals. If the monster ship, of rigid frame, were indeed the ideal form of dirigible it would be imitated on every hand. The inventions of the Wrights have been seized upon, adapted, improved perhaps by half a hundred aeroplane designers of every nation. But nobody has been imitating the Zeppelins.

That, however, is a mere passing reflection. If the Zeppelin has not done all in war that the sanguine German people expected of it, nevertheless it is not yet to be pronounced an entire failure. And even though a failure in war, the chief service for which its stout-hearted inventor designed it, there is still hope that it may ultimately prove better adapted to many ends of peace than the aeroplanes which for the time seem to have outdone it.

Stout-hearted indeed the old *Luftgraaf*—"Air Scout"—as the Germans call him, was. His was a Bismarckian nature, reminiscent of the Iron Chancellor alike physically and mentally. In appearance he recalls irresistibly the heroic figure of Bismarck, jack-booted and cuirassed at the Congress of Vienna, painted by von Werner. Heir to an old land-owning family, ennobled and entitled to bear the title *Landgraf*, Count von Zeppelin was a type of the German aristocrat. But for his title and aristocratic rank he could never have won his long fight for recognition by the bureaucrats who control the German army. In youth he was anti-Prussian in sentiment, and indeed some of his most interesting army experiences were in service with the army of South Germany against Prussia and her allied states. But all that was forgotten in the national unity that followed the defeat of France in 1872.

Before that, however, the young count—he was born in 1838—had served with gallantry, if not distinction, in the Union Army in our Civil War, had made a balloon ascension on the fighting line, had swum in the Niagara River below the falls, being rescued with difficulty, and together with two Russian officers and some Indian guides had almost starved in trying to discover the source of the Mississippi

The Giant and the Pigmies

River—a spot which can now be visited without undergoing more serious hardships than the upper berth in a Pullman car.

It was at the siege of Paris that Zeppelin's mind first became engaged with the problem of aerial navigation. From his post in the besieging trenches he saw the almost daily ascent of balloons in which mail was sent out, and persons who could pay the price sought to escape from the beleaguered city. As a colonel of cavalry, he had been employed mainly in scouting duty throughout the war. He was impressed now with the conviction that those globes, rising silently into the air, above the enemy's cannon shot and drifting away to safety would be the ideal scouts could they but return with their intelligence. Was there no way of guiding these ships in the air, as a ship in the ocean is guided? The young soldier was hardly home from the war when he began to study the problem. He studied it indeed so much to the exclusion of other military matters that in 1890 the General Staff abruptly dismissed him from his command. They saw no reason why a major-general of cavalry should be mooning around with balloons and kites like a schoolboy.

The dismissal hurt him, but deterred him in no way from the purpose of his life. Indeed the fruit of his many years' study of aeronautic conditions was ready for the gathering at this very moment. On the surface of the picturesque Lake Constance, on the border line between Germany and Switzerland, floated a huge shed, open to the water and more than five hundred feet long. In it, nearing completion, floated the first Zeppelin airship.

In the long patient study which the count had given to his problem he had reached the fixed conclusion that the basis of a practical dirigible balloon must be a rigid frame over which the envelope should be stretched. His experiments were made at the same time as those of Santos-Dumont, and he could not be ignorant of the measure of success which the younger man was attaining with the non-rigid balloon. But it was a fact that all the serious accidents which befell Santos-Dumont and most of the threatened accidents which he narrowly escaped were fundamentally caused by the lack of rigidity in his balloon. The immediate cause may have been a leaky valve permitting the gas to escape, or a faulty air-pump which made prompt filling of the ballonet impossible. But the effect of these flaws was to deprive the balloon of its rigidity, cause it to buckle, throwing the cordage out of gear, shifting stresses and strains, and resulting in ultimate breakdown.

Whether he observed the vicissitudes of his rival or not, Count

Zeppelin determined that the advantages of a rigid frame counted for more than the disadvantage of its weight. Moreover that disadvantage could be compensated for by increasing the size, and therefore the lifting power of the balloon. In determining upon a rigid frame the count was not a pioneer even in his own country. While his experiments were still under way, a rival, David Schwartz, who had begun, without completing, an airship in St. Petersburg, secured in some way aid from the German Government, which was at the moment coldly repulsing Zeppelin. He planned and built an aluminium airship but died before its completion. His widow continued the work amidst constant opposition from the builders. The end was one of the many tragedies of invention. Nobody but the widow ever believed the ship would rise from its moorings. It was in charge of a man who had never made an ascent. To his amazement and to the amazement of the spectators the engine was hardly started when the ship mounted and made headway against a stiff breeze. On the ground the spectators shouted in wonder; the widow, overwhelmed by this reward for her faith in her husband's genius, burst into tears of joy. But the amateur pilot was no match for the situation. Affrighted to find himself in mid-air, too dazed to know what to do,

A FRENCH "SAUSAGE"

he pulled the wrong levers and the machine crashed to earth. The pilot escaped, but the airship which had taken four years to build was irretrievably wrecked. The widow's hopes were blasted, and the way was left free for the Count von Zeppelin.

Freed, though unwillingly, from the routine duties of his military rank, Zeppelin thereafter devoted himself wholly to his airships. He was fifty-three years old, adding one more to the long list of men who found their real life's work after middle age. With him was associated his brother Eberhard, the two forming a partnership in aeronautical work as inseparable as that of Wilbur and Orville Wright. Like Wilbur Wright, Eberhard von Zeppelin did not live to witness the fullest fruition of the work, though he did see the soundness of its principles thoroughly established and in practical application. There is a picturesque story that when Eberhard lay on his death-bed his brother, instead of watching by his side, took the then completed airship from its hangar, and drove it over and around the house that the last sounds to reach the ears of his faithful ally might be the roar of the propellers in the air—the grand paean of victory.

Though Count von Zeppelin had begun his experiments in 1873 it was not until 1890 that he actually began the construction of his first airship. The intervening years had been spent in constructing and testing models, in abstruse calculations of the resistance of the air, the lifting power of hydrogen, the comparative rigidity and weight of different woods and various metals, the power and weight of the different makes of motors. In these studies he spent both his time and his money lavishly, with the result that when he had built a model on the lines of which he was willing to risk the construction of an airship of operative size, his private fortune was gone. It is the common lot of inventors. For a time the count suffered all the mortification and ignominy which the beggar, even in a most worthy cause, must always experience. Hat in hand he approached every possible patron with his story of certain success if only supplied with funds with which to complete his ship. A stock company with a capital of $225,000 of which he contributed one half, soon found its resources exhausted and retired from the speculation. Appeals to the Emperor met with only cold indifference. An American millionaire newspaper owner, resident in Europe, sent contemptuous word by his secretary that he "had no time to bother with crazy inventors." That was indeed the attitude of the business classes at the moment when the inventors of dirigibles were on the very point of conquering the obstacles in the

A British "blimp"

way of making the navigation of air a practical art. A governmental commission at Berlin rejected with contempt the plans which Zeppelin presented in his appeal for support. Members of that commission were forced to an about-face later and became some of the inventor's sturdiest champions. But in his darkest hour the government failed him, and the one friendly hand stretched out in aid was that of the German Engineers' Society which, somewhat doubtfully, advanced some funds to keep the work in operation.

With this the construction of the first Zeppelin craft was begun. Though there had been built up to the opening of the war twenty-five "Zeps"—nobody knows how many since—the fundamental type was not materially altered in the later ones, and a description of the first will stand for all. In connection with this description may be noted the criticisms of experts some of which proved only too well founded.

The first Zeppelin was polygonal, 450 feet long, 78 broad, and 66 feet high. This colossal bulk, equivalent to that of a 7500-ton ship necessary to supply lifting power for the metallic frame, naturally made her unwieldy to handle, unsafe to leave at rest, outside of a sheltering shed, and a particularly attractive target for artillery in time of war. Actual action indeed proved that to be safe from the shells of anti-aircraft guns, the Zeppelins were forced to fly so high that their own bombs could not be dropped with any degree of accuracy upon a desired target.

The balloon's frame is made of aluminium, the lightest of metals, but not the least costly. A curious disadvantage of this construction was made apparent in the accident which destroyed *Zeppelin IV.* That was the first of the airships to be equipped with a full wireless outfit which was used freely on its flight. It appeared that the aluminium frame absorbed much of the electricity generated for the purpose of the wireless. The effect of this was two-fold. It limited the radius of operation of the wireless to 150 miles or less, and it made the metal frame a perilous storehouse of electricity. When *Zeppelin IV.* met with a disaster by a storm which dragged it from its moorings, the stored electricity in her frame was suddenly released by contact with the trees and set fire to the envelope, utterly destroying the ship.

The balloon frame was divided into seventeen compartments, each of which held a ballonet filled with hydrogen gas. The purpose of this was similar to the practice of dividing a ship's hulls into compartments. If one or more of the ballonets, for any reason, were injured the remainder would keep the ship afloat. The space between the ballonets and the outer skin was pumped full of air to keep the latter taut and rigid. Moreover it helped to prevent the radiation of heat to the gas bags from the outer envelope whose huge expanse, presented to the sun, absorbed an immense amount of heat rays.

Two cars were suspended from the frame of the Zeppelin, forward and aft, and a corridor connected them. A sliding weight was employed to raise or depress the bow. In each car of the first Zeppelin was a sixteen-horse-power gasoline motor, each working two screws, with four foot blades, revolving one thousand times a minute. The engines were reversible, thus making it possible to work the propellers against each other and aid materially in steering the ship. Rudders at bow and stern completed the navigating equipment.

In the first Zeppelins, the corridor connecting the two cars was wholly outside the frame and envelope of the car. Later the perilous experiment was tried of putting it within the envelope. This resulted in one of the most shocking of the many Zeppelin disasters. In the case of the ship *L-II.*, built in 1912, the corridor became filled with gas that had oozed out of the ballonets. At one end or the other of the corridor this gas, then mixed with air, came in contact with fire,—perhaps the exhaust of the engines,—a violent explosion followed while the ship was some nine hundred feet aloft, and the mass of twisted and broken metal, with the flaming envelope, fell to the ground carrying twenty-eight men, including members of the Admiralty Board, to a horrible death.

But to return to the first Zeppelin. Her trial was set for July 2, 1900, and though the immediate vicinity of the floating hangar was barred to the public by the military authorities, the shores and surface of the lake were black with people eager to witness the test. Boats pulled out of the wide portal the huge cigar-shaped structure, floating on small rafts, its polished surface of *pegamoid* glittering in the sun. As large as a fair-sized ocean steamship, it looked, on that little lake dotted with pleasure craft, like a leviathan. Men were busy in the cars, fore and aft. The mooring ropes were cast off as the vessel gained an offing, and ballast being thrown out she began to rise slowly. The propellers began to whir, and the great craft swung around breasting the breeze and moved slowly up the lake. The crowd cheered. Count von Zeppelin, tense with excitement, alert for every sign of weakness watched his monster creation with mingled pride and apprehension. Two points were set at rest in the first two minutes—the lifting power was great enough to carry the heaviest load ever imposed upon a balloon and the motive power was sufficient to propel her against an ordinary breeze. But she was hardly in mid-air when defects became apparent. The apparatus for controlling the balancing weight got out of order. The steering lines became entangled so that the ship was first obliged to stop, then by reversing the engines to proceed backwards. This was, however, a favourable evidence of her handiness under untoward circumstances. After she had been in the air nearly an hour and had covered four or five miles, a landing was ordered and she dropped to the surface of the lake with perfect ease. Before reaching her shed, however, she collided with a pile—an accident in no way attributable to her design—and seriously bent her frame.

The story told thus baldly does not sound like a record of glorious success. Nevertheless not Count Zeppelin alone but all Germany was wild with jubilation. *Zeppelin I.* had demonstrated a principle; all that remained was to develop and apply this principle and Germany would have a fleet of aerial dreadnoughts that would force any hostile nation to subjection. There was little or no discussion of the application of the principle to the ends of peace. It was as an engine of war alone that the airship appealed to the popular fancy.

But at the time that fancy proved fickle. With a few repairs the airship was brought out for another test. In the air it did all that was asked for it, but it came to earth—or rather to the surface of the lake—with a shock that put it out of commission. When Count Zeppelin's company estimated the cost of further repairs it gave a sigh and abandoned the wreck. Thereupon the pertinacious inventor laid aside his tools, got into his old uniform, and went out again on the dreary task of begging for further funds.

It was two years before he could take up again the work of construction. He lectured, wrote magazine articles, begged, cajoled, and pleaded for money. At last he made an impression upon the Emperor who, indeed, with a keen eye for all that makes for military advantage, should have given heed to his efforts long before. Merely a letter of approval from the all-powerful Kaiser was needed to turn the scale and in 1902 this was forthcoming. The factories of the empire agreed to furnish materials at cost price, and sufficient money was soon forthcoming to build a second ship. This ship took more than two years to build, was tested in January, 1906, made a creditable flight, and was dashed to pieces by a gale the same night!

The wearisome work of begging began again. But this time the Kaiser's aid was even more effectively given and in nine months *Zeppelin III.* was in the air. More powerful than its predecessors it met with a greater measure of success. On one of its trials a propeller blade flew off and penetrated the envelope, but the ship returned to earth in safety. In October, 1906, the Minister of War reported that the airship was extremely stable, responded readily to her helm, had carried eleven persons sixty-seven miles in two hours and seventeen minutes, and had made its landing in ease and safety. Accepted by the government "*No. III.*" passed into military service and Zeppelin, now the idol of the German people, began the construction of "*No. IV.*"

That ship was larger than her predecessors and carried a third cabin for passengers suspended amidships. Marked increase in the size of the steering and stabling planes characterized the appearance of the ship when compared with earlier types. She was at the outset a lucky ship. She cruised through Alpine passes into Switzerland, and made a circular voyage carrying eleven passengers and flying from Friedrichshaven to Mayence and back via Basle, Strassburg, Mannheim, and Stuttgart. The voyage occupied twenty-one hours—a world's record. The performance of the ship on both voyages was perfection. Even in the tortuous Alpine passes which she was forced to navigate on her trip to Lucerne she moved with the steadiness and certainty of a great ship at sea. The rarefication of the air at high altitudes, the extreme and sudden variations in temperature, the gusts of wind that poured from the ice-bound peaks down through the narrow canyons affected her not at all. When to this experience was added the triumphant tour of the six German cities, Count von Zeppelin might well have thought his triumph was complete.

But once again the cup of victory was dashed from his lips. After his landing a violent wind beat upon the ship. An army of men strove to hold her fast, while an effort was made to reduce her bulk by deflation. That effort, which would have been entirely successful in the case of a non-rigid balloon, was obviously futile in that of a Zeppelin. Not the gas in the ballonets, but the great rigid frame covered with waterproofed cloth constituted the huge bulk that made her the plaything of the winds. In a trice she was snatched from the hands of her crew and hurled against the trees in a neighbouring grove. There was a sudden and utterly unexpected explosion and the whole fabric was in flames. The precise cause of the explosion will always be in doubt, but, as already pointed out, many scientists believe that the great volume of electricity accumulated in the metallic frame was suddenly released in a mighty spark which set fire to the stores of gasoline on board.

With this disaster the iron nerve of the inventor was for the first time broken. It followed so fast upon what appeared to be a complete triumph that the shock was peculiarly hard to bear. It is said that he broke down and wept, and that but for the loving courage and earnest entreaties of his wife and daughter he would then have abandoned the hope and ambition of his life. But after all it was but that darkest hour which comes just before the dawn. The demolition of *"No. IV."* had been no accident which reflected at all upon the plan or construction of the craft—unless the great bulk of the ship be considered a fundamental defect. What it did demonstrate was that the Zeppelin, like the one-thousand-foot ocean liner, must have adequate harbour and docking facilities wherever it is to land. The one cannot safely drop down in any convenient meadow, any more than the other can put into any little fishing port. Germany has learned this lesson well enough and since the opening of the Great War her territory is plentifully provided with Zeppelin shelters at all strategic points.

Fortunately for the Count the German people judged his latest reverse more justly than he did. They saw the completeness of the triumph which had preceded the disaster and recognized that the latter was one easily guarded against in future. Enthusiasm ran high all over the land. Begging was no longer necessary. The Emperor, who had heretofore expressed rather guarded approval of the enterprise, now flung himself into it with that enthusiasm for which he is notable. He bestowed upon the Count the Order of the Black Eagle, embraced

THE DEATH OF A ZEPPELIN

him in public three times, and called aloud that all might hear, "Long life to his Excellency, Count Zeppelin, the Conqueror of the Air." He never wearied of assuring his hearers that the Count was the "greatest German of the century." With such august patronage the Count became the rage. Next to the Kaiser's the face best known to the people of Germany, through pictures and statues, was that of the inventor of the Zeppelin. The pleasing practice of showing affection for a public man by driving nails into his wooden effigy had not then been invented by the poetic Teutons, else von Zeppelin would have outdone von Hindenburg in weight of metal.

The story that Zeppelin had refused repeated offers from other governments was widely published and evoked patriotic enthusiasm. With it went shrewd hints that in these powerful aircraft lay the way to overcome the hated English navy, and even to carry war to the very soil of England. It was then eight years before the greatest war of history was to break out, but even at that date hatred of England was being sedulously cultivated among the German people by those in authority.

As a result of this national attitude Count Zeppelin's enterprise was speedily put on a sound financial footing. Though *"No. IV."* had been destroyed by an accident it had been the purpose of the government to buy her, and $125,000 of the purchase price was now put at the disposal of the Count von Zeppelin. A popular Zeppelin fund of $1,500,000 was raised and expended in building great works. Thenceforward there was no lack of money for furthering what had truly become a great national interest.

But the progress of the construction of Zeppelins for the next few years was curiously compounded of success and failure. Fate seemed to have decreed to every Zeppelin triumph a disaster. Each mischance was attributed to exceptional conditions which never could happen again, but either they did occur, or some new but equally effective accident did. Outside of Germany, where the public mind had become set in an almost idolatrous confidence in Zeppelin, the great airships were becoming a jest and a byword notwithstanding their unquestioned accomplishments. Indeed when the record was made up just before the declaration of war in 1914 it was found that of twenty-five Zeppelins thus far constructed only twelve were available. Thirteen had been destroyed by accident—two of them modern naval airships only completed in 1913. The record was not one to inspire confidence.

In 1909, during a voyage in which he made nine hundred miles in thirty-eight hours, the rumour was spread that von Zeppelin would continue it to Berlin. Some joker sent a forged telegram to the Kaiser to that effect signed "Zeppelin." It was expected to be the first appearance of one of the great ships at the capital, and the Emperor hastened to prepare a suitable welcome. A great crowd assembled at the Templehoff Parade Ground. The Berlin Airship Battalion was under orders to assist in the landing. The Kaiser himself was ready to hasten to the spot should the ship be sighted. But she never appeared. If von Zeppelin knew of the exploit which rumour had assigned to him—which is doubtful—he could not have carried it out. His ship collided with a tree—an accident singularly frequent in the Zeppelin records—so disabling it that it could only limp home under half power. A rather curt telegram from his Imperial master is said to have been Count von Zeppelin's first intimation that he had broken an engagement.

However, he kept it two months later, flying to Berlin, a distance of 475 miles. He was greeted with mad enthusiasm and among the crowd to welcome him was Orville Wright the American aviator. It is a curious coincidence that on the day the writer pens these words the New York newspapers contain accounts of Mr. Wright's proffer of his services, and aeronautical facilities, to the President in case an existing diplomatic break with Germany should reach the point of actual war. Mr. Wright accompanied his proffer by an appeal for a tremendous aviation force, "but," said he, "I strongly advise against spending any money whatsoever on dirigible balloons of any sort."

Thereafter the progress of Count von Zeppelin was without interruption for any lack of financial strength. His great works at Friedrichshaven expanded until they were capable of putting out a complete ship in eight weeks. He was building, of course, primarily for war, and never concealed the fact that the enemy he expected to be the target of his bomb throwers was England. What the airships accomplished in this direction, how greatly they were developed, and the strength and weakness of the German air fleet, will be dwelt upon in another chapter.

But, though building primarily for military purposes, Zeppelin did not wholly neglect the possibilities of his ship for non-military service. He built one which made more than thirty trips between Munich and Berlin, carrying passengers who paid a heavy fee for the privilege of enjoying this novel form of travel. The car was fitted up like our most up-to-date Pullmans, with comfortable seats, bright lights, and a kitchen from which excellent meals were served to the passengers. The service was not continued long enough to determine

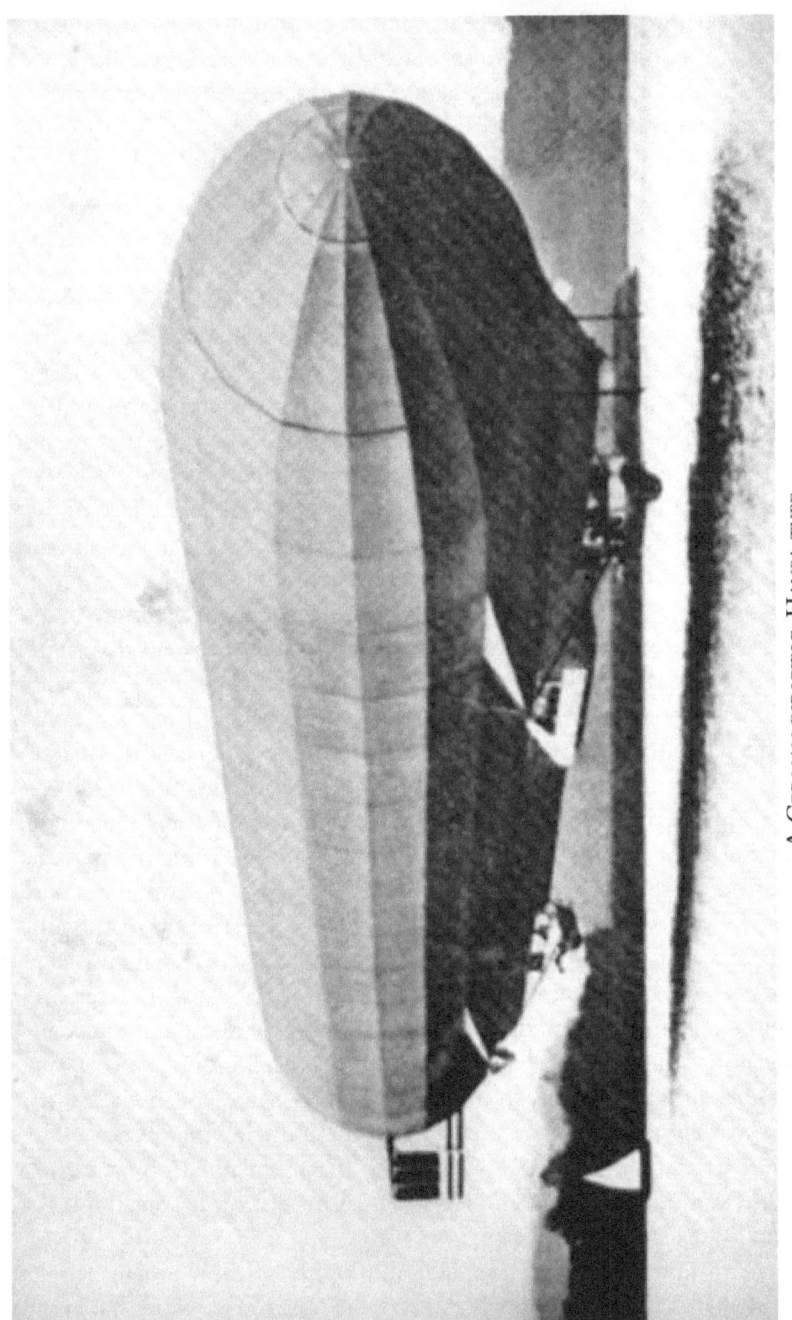

A German dirigible, Hansa type

whether it could ever be made commercially profitable, but as an aid to firing the Teutonic heart and an assistance in selling stock it was well worth while. The spectacle of one of these great cars, six hundred or more feet long, floating grandly on even keel and with a steady course above one of the compact little towns of South Germany, was one to thrill the pulses.

But the ill luck which pursued Count von Zeppelin even in what seemed to be his moments of assured success was remorseless. In 1912 he produced the monster *L-I*, 525 feet long, 50 feet in diameter, of 776,900 cubic feet capacity, and equipped with three sets of motors, giving it a speed of fifty-two miles an hour. This ship was designed for naval use and after several successful cross-country voyages she was ordered to Heligoland, to participate in naval manoeuvres with the fleet there stationed. One day, caught by a sudden gust of wind such as are common enough on the North Sea, she proved utterly helpless. Why no man could tell, her commander being drowned, but in the face of the gale she lost all control, was buffeted by the elements at their will, and dropped into the sea where she was a total loss. Fifteen of her twenty-two officers and men were drowned. The accident was the more inexplicable because the craft had been flying steadily overland for nearly twelve months and had covered more miles than any ship of Zeppelin construction. It was reported that her captain had said she was overloaded and that he feared that she would be helpless in a gale. But after the disaster his mouth was stopped by the waters of the North Sea.

This calamity was not permitted long to stand alone. Indeed one of the most curious facts about the Zeppelin record is the regular, periodical recurrence of fatal accidents at almost equal intervals and apparently wholly unaffected by the growing perfection of the airships. While *L-I* was making her successful cross-country flights, *L-II* was reaching completion at Friedrichshaven. She was shorter but bulkier than her immediate predecessor and carried engines giving her nine hundred horse power, or four hundred more than *L-I*. On its first official trip this ship exploded a thousand feet in air, killing twenty-eight officers and men aboard, including all the officials who were conducting the trials. The calamity, as explained on an earlier page, was due to the accumulation of gas in the communicating passage between the three cars.

This new disaster left the faith and loyalty of the German people unshaken. But it did decidedly estrange the scientific world from Count von Zeppelin and all his works. It was pointed out, with truth, that the

A WRECKED ZEPPELIN AT SALONIKA

accident paralleled precisely one which had demolished the *Severo Pax* airship ten years earlier, and which had caused French inventors to establish a hard and fast rule against incorporating in an airship's design any enclosed space in which waste gas might gather. This rule and its reason were known to Count von Zeppelin and by ignoring both he lent new colour to the charge, already current in scientific circles, that he was loath to profit by the experiences of other inventors.

Whether this feeling spread to the German Government it is impossible to say. Nor it is easy to estimate how much official confidence was shaken by it. The government, even before the war, was singularly reticent about the Zeppelins, their numbers and plans. It is certain that orders were not withheld from the count. Great numbers of his machines were built, especially after the war was entered upon. But he was not permitted longer to have a monopoly of government aid for manufacturers of dirigibles. Other types sprung up, notably the Schutte-Lanz, the Gross, and the Parseval. But being first in the field the Zeppelin came to give its name to all the dirigibles of German make and many of the famous—or infamous—exploits credited to it during the war may in fact have been performed by one of its rivals.

It would be futile to attempt to enumerate all these rivals here. Among them are the semi-rigid Parseval and Gross types which found great favour among the military authorities during the war. The latter is merely an adaptation of the highly successful French ship the *Lebaudy*, but the Parseval is the result of a slow evolution from an ordinary balloon. It is wholly German, in conception and development, and it is reported that the Kaiser, secretly disgusted that the Zeppelins, to the advancement of which he had given such powerful aid, should have recorded so many disasters, quietly transferred his interest to the new and simpler model. Despite the hope of a more efficient craft, however, both the Gross and the Parseval failed in their first official trials, though later they made good.

The latter ship was absolutely without any wooden or metallic structure to give her rigidity. Two air ballonets were contained in the envelope at bow and stern and the ascent and descent of the ship was regulated by the quantity of air pumped into these. A most curious device was the utilization of heavy cloth for the propeller blades. Limp and flaccid when at rest, heavy weights in the hem of the cloth caused these blades to stand out stiff and rigid as the result of the centrifugal force created by their rapid revolution. One great military advantage of the Parseval was that she could be quickly deflated in the presence of danger at her moorings, and wholly knocked down and packed in small compass for shipment by rail in case of need. To neither of these models did there ever come such a succession of disasters as befell the earlier Zeppelins. It is fair to say however that prior to the war not many of them had been built, and that both their builders and navigators had opportunity to learn from Count von Zeppelin's errors.

Among the chief German rivals to the Zeppelin is the Schutte-Lanz, of the rigid type, broader but not so long as the Zeppelin, framed of wood bound with wire and planned to carry a load of five or six tons, or as many as thirty passengers. *No. I* of this type met its fate as did so many Zeppelins by encountering a storm while improperly moored. Called to earth to replenish its supply of gas it was moored to an anchor sunk six feet in the ground, and as an additional precaution three hundred soldiers were called from a neighbouring barracks to handle it. It seems to have been one of the advantages of Germany as a place in which to manoeuvre dirigibles, that, even in time of peace, there were always several hundred soldiers available wherever a ship might land. But this force was inadequate. A violent gust tore the ship from their hands. One poor fellow instinctively clung to his rope until one thousand feet in the air when he let go. The ship itself hovered over the town for an hour or more, then descended and was dashed to pieces against trees and stone walls.

The danger which was always attached to the landing of airships has led some to suggest that they should never be brought to earth, but moored in mid-air as large ships anchor in midstream. It is suggested that tall towers be built to the top of which the ship be attached by a cable, so arranged that she will always float to the leeward of the tower. The passengers would be landed by gangplanks, and taken up and down the towers in elevators. Kipling suggests this expedient in his prophetic sketch *With the Night Mail*. The airship would only return to earth—as a ship goes into dry dock—when in need of repairs.

A curious mishap that threatened for a time to wreck the peace of the world, occurred in April, 1913, when a German Zeppelin was forced out of its course and over French territory. The right of alien machines to pass over their territory is jealously guarded by European nations, and during the progress of the Great War the Dutch repeatedly protested against the violation of their atmosphere by German aviators. At the time of this mischance, however, France and Germany were at peace—or as nearly so as racial and historic antipathies would permit. Accordingly when officers of a brigade of French cavalry en-

BRITISH AVIATORS ABOUT TO ASCEND:
NOTE POSITION OF GUNNER ON LOWER SEAT

gaged in manoeuvring near the great fortress of Luneville saw a shadow moving across the field and looking up saw a huge Zeppelin betwixt themselves and the sun they were astonished and alarmed. Signs and faint shouts from the aeronauts appeared to indicate that their errand was at least friendly, if not involuntary. The soldiers stopped their drill; the townspeople trooped out to the Champs de Mars where the phenomenon was exhibited and began excitedly discussing this suspicious invasion. Word was speedily sent to military headquarters asking whether to welcome or to repel the foe.

Meantime the great ship was drifting perilously near the housetops, and the uniformed officers in the cars began making signals to the soldiers below. Ropes were thrown out, seized by willing hands and made fast. The crew of Germans descended to find themselves prisoners. The international law was clear enough. The ship was a military engine of the German army. Its officers, all in uniform, had deliberately steered her into the very heart of a French fortress. Though the countries were at peace the act was technically one of war—an armed invasion by the enemy. Diplomacy of course settled the issue peacefully but not before the French had made careful drawings of all the essential features of the Zeppelin, and taken copies of its log. As Germany had theretofore kept a rigid secrecy about all the details of Zeppelin construction and operation this angered the military authorities beyond measure. The unlucky officers who had shared in the accident were savagely told that they should have blown the ship up in mid-air and perished with it rather than to have weakly submitted it to French inspection. They suffered court-martial but escaped with severe reprimands.

The story of the dirigibles of France and Germany is practically the whole story of the development to a reasonable degree of perfection of the lighter-than-air machine. Other nations experimented somewhat, but in the main lagged behind these pioneers. Out of Spain indeed came a most efficient craft—the Astra-Torres, of which the British Government had the best example prior to the war, while both France and Russia placed large orders with the builders. How many finally went into service and what may have been their record are facts veiled in the secrecy of wartime. Belgium and Italy both produced dirigibles of distinctive character. The United States is alone at the present moment in having contributed nothing to the improvement of the dirigible balloon.

CHAPTER 5

The Development of the Aeroplane

The story of the development of the heavier-than-air machine—which were called aeroplanes at first, but have been given the simpler name of aeroplanes—is far shorter than that of the balloons. It is really a record of achievement made since 1903 when the plane built by Professor Langley of the Smithsonian Institution came to utter disaster on the Potomac. In 1917, at the time of writing this book, there are probably thirty distinct types of aeroplanes being manufactured for commercial and military use, and not less than fifty thousand are being used daily over the battlefields of Europe. No invention save possibly the telephone and the automobile ever attained so prodigious a development in so brief a time. Wise observers hold that the demand for these machines is yet in its infancy, and that when the end of the war shall lead manufacturers and designers to turn their attention to the commercial value of the aeroplane the flying craft will be as common in the air as the automobiles at least on our country roads.

The idea of flying like a bird with wings, the idea basically underlying the aeroplane theory, is old enough—almost as old as the first conception of the balloon, before hydrogen gas was discovered. In an earlier chapter some account is given of early experiments with wings. No progress was made along this line until the hallucination that man could make any headway whatsoever against gravity by flapping artificial wings was definitely abandoned. There was more promise in the experiments made by Sir George Cayley, and he was followed in the first half of the nineteenth century by half a dozen British experimenters who were convinced that a series of planes, presenting a fixed angle to the breeze and driven against it by a sufficiently powerful motor, would develop a considerable lifting power. This was demonstrated by Henson, in 1842, Stringfellow, in 1847, Wenham, who arranged his planes like slats in a Venetian blind and first applied the modern term "aeroplane" to his invention, and Sir Hiram Maxim, who built in 1890 the most complicated and impressive looking 'plane the world has yet seen. But though each of these inventors proved the theorem that a heavier-than-air machine could be made to fly, all failed to get practical results because no motor had then been invented which combined the necessary lightness with the generation of the required power.

In America we like to think of the brothers Wright as being the true inventors of the aeroplane. And indeed they did first bring it to the point of usefulness, and alone among the many pioneers lived to see the adoption of their device by many nations for serious practical use. But it would be unjust to claim for them entire priority in the field of the glider and the heavier-than-air machine. Professor Langley preceded them with an aeroplane which, dismissed with ridicule as a failure in his day, was long after his death equipped with a lighter motor and flown by Glenn Curtis, who declared that the scientist had solved the problem, had only the explosive engine been perfected in his time.

Despite, however, the early period of the successful experiments of the Wrights and Professor Langley, it would be unjust for America to arrogate to herself entire priority in aeroplane invention. Any story of that achievement which leaves out Lilienthal, the German, and Pilcher, the Englishman, is a record in which the truth is subordinated to national pride.

Otto Lilienthal and his brother Gustav—the two like the Wrights were always associated in their aviation work—had been studying long the problem of flight when in 1889 they jointly published their book *Bird Flight as the Basis of the Flying Art*. Their investigations were wholly into the problem of flight without a motor. At the outset they even harked back to the long-abandoned theory that man could raise himself by mere muscular effort, and Otto spent many hours suspended at the end of a rope flapping frantically a pair of wings before he abandoned this effort as futile. Convinced that the soaring or gliding of the birds was the feat to emulate, he made himself a pair of fixed, bat-like wings formed of a light fabric stretched over a willow frame. A tail composed of one vertical and one horizontal plane extended to the rear, and in the middle the aviator hung by his armpits, in an erect position. With this device he made some experimental glides, leaping from slight eminences. With his body, which swung at will from its cushioned supports, he could balance, and even steer the fabric which

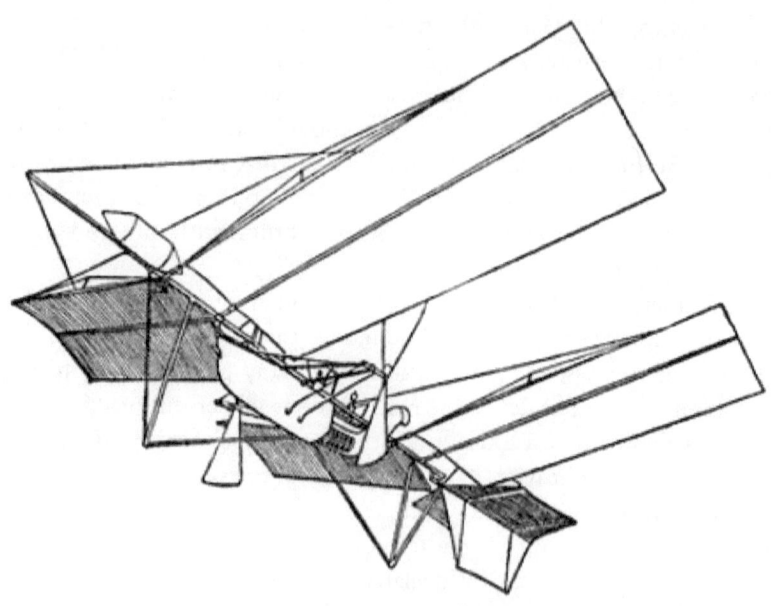

LANGLEY'S AEROPLANE

supported him, and accomplished long glides against the wind. Not infrequently, running into the teeth of the breeze down a gentle slope he would find himself gently wafted into the air and would make flights of as much as three hundred yards, steering to either side, or rising and falling at will. He was even able to make a circuitous flight and return to his starting place—a feat that was not accomplished with a motor-driven aeroplane until years later. Lilienthal achieved it with no mechanical aid, except the wings. He became passionately devoted to the art, made more than two thousand flights, and at the time of his death had just completed a motor-driven aeroplane, which he was never able to test. His earlier gliding wings he developed into a form of biplane, with which he made several successful flights, but met his death in 1896 by the collapse of this machine, of the bad condition of which he had been warned.

Lilienthal was more of a factor in the conquest of the air than his actual accomplishments would imply. His persistent experiments, his voluminous writings, and above all his friendly and intelligent interest in the work of other and younger men won him a host of disciples in other lands who took up the work that dropped from his lifeless hands.

In England Percy S. Pilcher emulated the Lilienthal glides, and was

FRENCH AERODROME NEAR THE FRONT

at work on a motor-propelled machine when he was killed by the breakage of a seemingly unimportant part of his machine. He was on the edge of the greater success, not to that moment attained by anyone, of building a true aeroplane propelled by motor. Many historians think that to Lilienthal and Pilcher is justly due the title "the first flying men." But Le Bris, a French sailor, utterly without scientific or technical equipment, as far back as 1854 had accomplished a wonderful feat in that line. While on a cruise he had watched an albatross that followed his ship day after day apparently without rest and equally without fatigue. His imagination was fired by the spectacle and probably having never heard of the punishment that befell the Ancient Mariner, he shot the albatross. "I took the wing," he wrote later, "and exposed it to the breeze, and lo, in spite of me, it drew forward into the wind; notwithstanding my resistance it tended to rise. Thus I had discovered the secret of the bird. I comprehend the whole mystery of flight."

A trifle too sanguine was sailor Le Bris, but he had just the qualities of imagination and confidence essential to one who sets forth to conquer the air. Had he possessed the accurate mind, the patience, and the pertinacity of the Wrights he might have beaten them by half a century. As it was he accomplished a remarkable feat, though it ended in somewhat laughable failure. He built an artificial bird, on the general plan of his albatross. The wings were not to flap, but their angles to the wind were controlled by a system of levers controlled by Le Bris, who stood up in the basket in the centre. To rise he required something like the flying start which the aeroplanes of to-day get on their bicycle wheels before leaving the ground. As Le Bris had no motor this method of propulsion was denied him, so he loaded the apparatus in a cart, and fastened it to the rail by a rope knotted in a slip knot which a jerk from him would release. As they started men walked

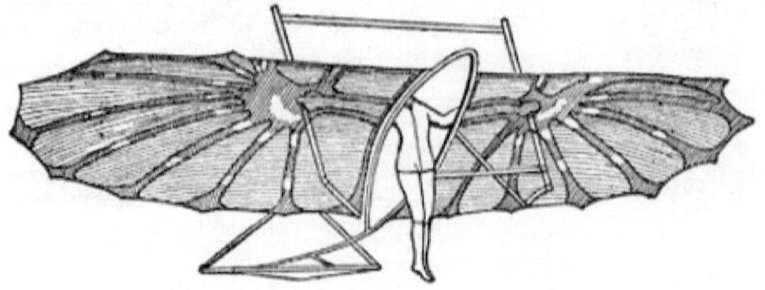

LILIENTHAL'S GLIDER

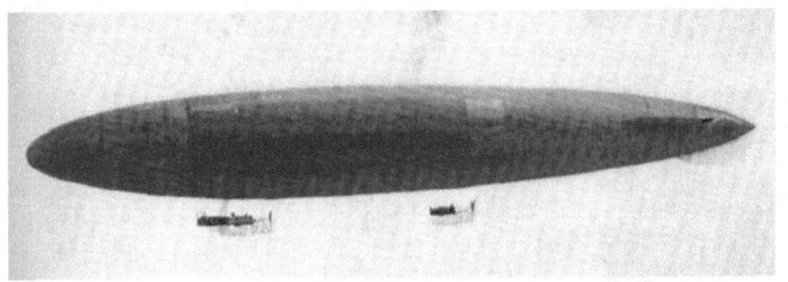

A GERMAN WAR ZEPPELIN

beside the cart holding the wings, which extended for twenty-five feet on either side. As the horses speeded up these assistants released their hold. Feeling the car try to rise under his feet Le Bris cast off the rope, tilted the front end of the machine, and to his joy began to rise steadily into the air. The spectators below cheered madly, but a note of alarm mingled with their cheers, and the untried aviator noticed a strange and inexplicable jerking of his machine. Peering down he discovered, to his amaze, a man kicking and crying aloud in deadly fear. It was evident that the rope he had detached from the cart had caught up the driver, who had thus become, to his intense dismay, a partner in the inventor's triumph. Indeed it is most possible that he contributed to that triumph for the ease and steadiness with which the machine rose to a height estimated at three hundred feet suggests that he may have furnished needed ballast—acted in fact as the tail to the kite. Humanity naturally impelled Le Bris to descend at once, which he did skilfully without injuring his involuntary passenger, and only slightly breaking one of the wings.

Had Le Bris won this success twenty years later his fame and fortune would have been secure. But in 1854 the time was not ripe for aeronautics. Le Bris was poor. The public responded but grudgingly to his appeals for aid. His next experiment was less successful—perhaps for lack of the carter—and he ultimately disappeared from aviation to become an excellent soldier of France.

Perhaps had they not met with early and violent deaths, the Lilienthals and Pilcher might have carried their experiments in the art of gliding into the broader domain of power flight. This however was left to the two Americans, Orville and Wilbur Wright, who have done more to advance the art of navigating the air than all the other experimenters whose names we have used. The story of the Wright brothers is one of boyhood interest gradually developed into the passion of a lifetime. It parallels to some degree the story of Santos-

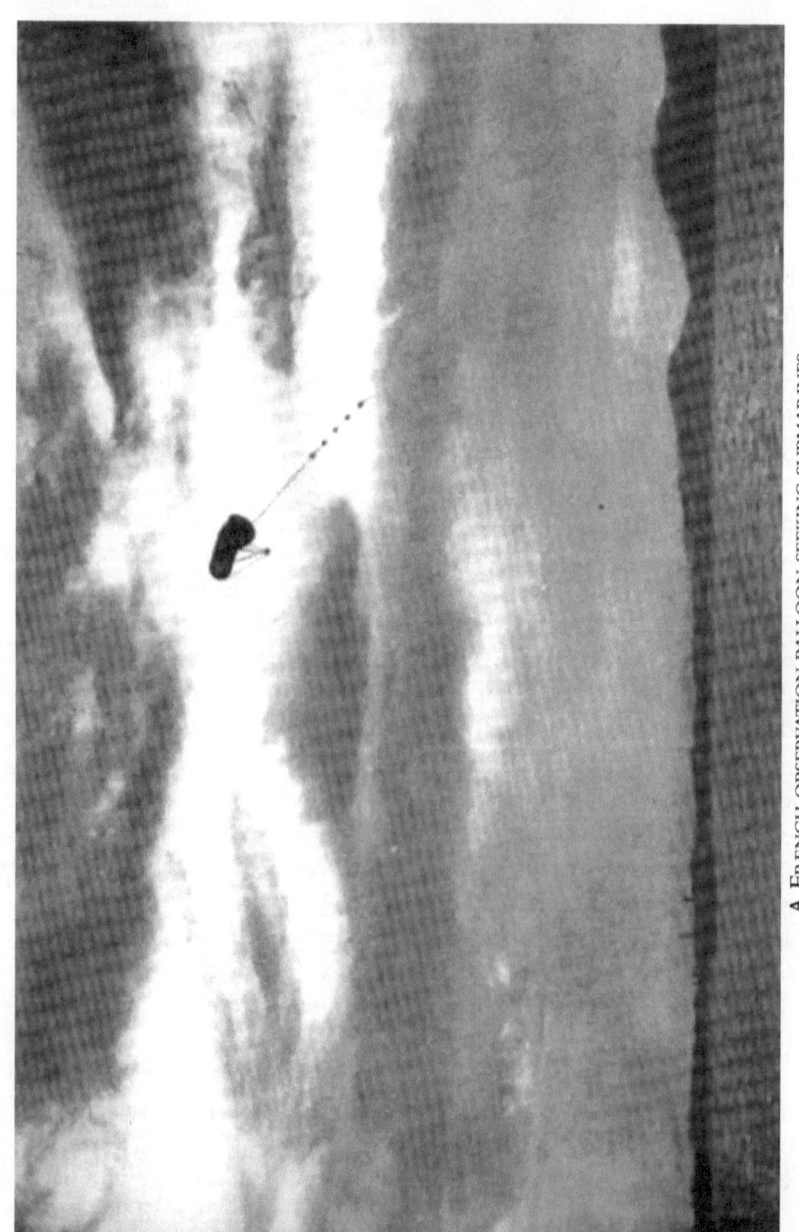

A French observation balloon seeking submarines

Dumont who insisting as a child that "man flies" finally made it a fact. The interest of the Wrights was first stimulated when, in 1878, their father brought home a small toy, called a "helicopter," which when tossed in the air rose up instead of falling. Every child had them at that time, but curiously this one was like the seed which fell upon fertile soil. The boys went mad, as boys will, on the subject of flying. But unlike most boys they nurtured and cultivated the passion and it stayed with them to manhood. From helicopters they passed to kites, and from kites to gliders. By calling they were makers and repairers of bicycles, but their spare time was for years devoted to solving the problem of flight. In time it became their sole occupation and by it they won a fortune and world-wide fame. Their story forms a remarkable testimony to the part of imagination, pertinacity, and courage in winning success. After years of tests with models, and with kites controlled from the ground, the brothers had worked out a type of glider which they believed, in a wind of from eighteen to twenty miles an hour, would lift and carry a man. But they had to find a testing ground. The fields near their home in Ohio were too level, and their firm unyielding surface was not attractive as a cushion on which to light in the event of disaster. Moreover the people round about were getting inquisitive about these grown men "fooling around" with kites and flying toys. To the last the Wrights were noted for their dislike of publicity, and it is entirely probable that the sneering criticisms of their "level headed" and "practical" neighbours had a good deal to do with rooting them in this distaste.

Low steep hills down the sides of which they could run and at the proper moment throw themselves upon their glider; a sandy soil which would at least lessen the shock of a tumble; and a vicinage in which winds of eighteen miles an hour or more is the normal atmospheric state were the conditions they sought. These they found at a little hamlet called Kitty-Hawk on the coast of North Carolina. There for uncounted centuries the tossing Atlantic had been throwing up its snowy sand upon the shore, and the steady wind had caught it up, piled it in windrows, rolled it up into towering hills, or carried it over into the dunes which extended far inland. It was a lonely spot, and there secure from observation the Wrights pitched their camp. For them it was a midsummer's holiday. Not at first did they decide to make aviation not a sport but a profession. To their camp came visitors interested in the same study, among them Chanute, a well-known experimenter, and some of his associates. They had thought to give hours at a time

CHANUTE'S GLIDER

to actual flight. When they closed their first season, they found that all their time spent in actual flight footed up less than an hour. Lilienthal, despite all he accomplished, estimated that he, up to a short time before his death, spent only about five hours actually in the air. In that early day of experimentation a glide covering one hundred feet, and consuming eight or ten seconds, was counted a triumph.

But the season was by no means wasted. Indeed such was the estimate that the Wrights put upon it that they folded their tents determined that when they returned the year following it would be as professionals, not amateurs. They were confident of their ability to build machines that would fly, though up to that time they had never mounted a motor on their aircraft.

In the clear hot air of a North Carolina midsummer the Wrights used to lie on their backs studying through glasses the methods of flight of the great buzzards—filthy scavenger birds which none the less soaring high aloft against a blue sky are pictures of dignity and grace.

> Bald eagles, ospreys, hawks, and buzzards give us daily exhibitions of their powers (wrote Wilbur Wright). The buzzards were the most numerous, and were the most persistent soarers.

They apparently never flapped except when it was absolutely necessary, while the eagles and hawks usually soared only when they were at leisure. Two methods of soaring were employed. When the weather was cold and damp and the wind strong the buzzards would be seen soaring back and forth along the hills or at the edge of a clump of trees. They were evidently taking advantage of the current of air flowing upward over these obstructions. On such days they were often utterly unable to soar, except in these special places. But on warm clear days when the wind was light they would be seen high in the air soaring in great circles. Usually, however, it seemed to be necessary to reach a height of several hundred feet by flapping before this style of soaring became possible. Frequently a great number of them would begin circling in one spot, rising together higher and higher till finally they would disperse, each gliding off in whatever direction it wished to go. At such times other buzzards only a short distance away found it necessary to flap frequently in order to maintain themselves. But when they reached a point beneath the circling flock they began to rise on motionless wings. This seemed to indicate that rising columns of air do not exist everywhere, but that the birds must find them. They evidently watch each other and when one finds a rising current the others quickly make their way to it. One day when scarce a breath of wind was stirring on the ground we noticed two bald eagles sailing in circling sweeps at a height of probably five hundred feet. After a time our attention was attracted to the flashing of some object considerably lower down. Examination with a field-glass proved it to be a feather which one of the birds had evidently cast. As it seemed apparent that it would come to earth only a short distance away, some of our party started to get it. But in a little while it was noted that the feather was no longer falling, but on the contrary was rising rapidly. It finally went out of sight upward. It apparently was drawn into the same current in which the eagles were soaring and was carried up like the birds.

It was by such painstaking methods as these, coupled with the mathematical reduction of the fruits of such observations to terms of angles and supporting planes, that the Wrights gradually perfected their machine. The first aeroplane to which they fitted a motor and which actu-

A German Taube pursued by British planes

ally flew has been widely exhibited in the United States, and is to find final repose in some public museum. Study it as you will you can find little resemblance in those rectangular rigid planes to the wings of a bird. But it was built according to deductions drawn from natural flight.

The method of progress in these preliminary experiments was, by repeated tests, to determine what form of aeroplane, and of what proportions, would best support a man. It was evident that for free and continuous flight it must be able to carry not only the pilot, but an engine and a store of fuel as well. Having, as they thought, determined these conditions the Wrights essayed their first flight at their home near Dayton, Ohio. It was a cold December day in 1903. The first flight, with motor and all, lasted twelve seconds; the fourth fifty-nine seconds. The handful of people who came out to witness the marvel went home jeering. In the spring of the next year a new flight was announced near Dayton. The newspapers had been asked to send reporters. A crowd of perhaps fifty persons had gathered. Again fate was hostile. The engine worked badly and the aeroplane refused to rise. The crowd dispersed and the newspapermen, returning the next day, met only with another disappointment.

These repeated failures in public exhibitions resulted in creating general indifference to the real progress that the Wrights were making in solving the flight problem. While the gliding experiments at Kitty-Hawk were furnishing the data for the plans on which the tens of thousands of aeroplanes used in the European war were afterwards built, no American newspaper was sufficiently interested to send representatives to the spot. The people of the United States were supremely indifferent. Perhaps this was due to the fact that superficially

THE FIRST WRIGHT GLIDER

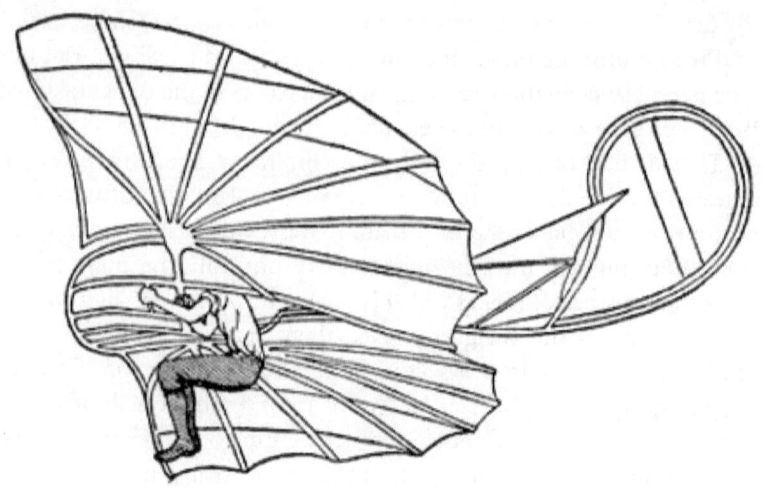

PILCHER'S GLIDER

regarded the machine the Wrights were trying to perfect gave promise of usefulness only in war or in sport. We are not either a warlike or a sporting people. Ready enough to adopt a new device which seems adapted for utilitarian purposes, as is shown by the rapid multiplication of automobiles, we leave sport to our professional ball players, and our military equipment to luck.

So after continued experimental flights in the open fields near Dayton had convinced them that the practical weaknesses in their machine had been eliminated, the Wrights packed up their flyer and went to France. Before so doing they tried to get encouragement from the United States Government, but failed. Neither the government nor any rich American was willing to share the cost of further experiments. All that had been done was at their own cost, both in time and money. In France, whither they went in 1908, they had no coldness to complain of. It was then the golden day of aviation in the land which always afforded to the Knights of the Air their warmest welcome and their most liberal support. Two years had elapsed since Santos-Dumont, turning from dirigibles to 'planes, had made a flight of 238 yards. This the Wrights had at the time excelled at home but without attracting attention. France on the contrary went mad with enthusiasm, and claimed for the Brazilian the honour of first demonstrating the possibility of flight in a heavier-than-air machine. England, like the United States, was cold, clinging to the balloon long after all other nations had abandoned it. But France welcomed the

THE COMPARATIVE STRENGTH OF BELLIGERENTS IN AEROPLANES AT THE OPENING OF THE WAR.
The French Army had at least 500 aeroplanes. England had about 250 aeroplanes of all types Russia had 50 aeroplanes—Austria had at least 50 aeroplanes Germany is about the equal of France, having 500 flyers.

Wrights with enthusiasm. They found rivals a-plenty in their field of effort. Santos-Dumont, Bleriot, Farman, Latham were all flying with aeroplanes, but with models radically different from that of the American brothers. Nevertheless the latter made an instant success.

From the moment they found that they had hit upon the secret of raising, supporting, and propelling an aeroplane, the Wrights made of their profession a matter of cold business. In many ways this was the best contribution they could possibly have made to the science of aviation, though their keen eye to the main chance did bring down on them a certain amount of ridicule. Europe laughed long at the *sang-froid* with which Wilbur Wright, having won the Michelin prize of eight hundred pounds, gave no heed to the applause which the assembled throng gave him as the money was transferred to him with a neat presentation speech. Without a word he divided the notes into two packets, handed one to his brother Orville, and thrust the other into his own pocket. For the glory which attended his achievement he cared nothing. It was all in the day's work. Later in the course of trials of a machine for the United States Government at Fort Myer, just across the Potomac from Washington, the Wrights seriously offended a certain sort of public sentiment in a way which undoubtedly set back the encouragement of aviation by the United States Government very seriously.

In 1909, they had received a contract from the government for a machine for the use of the Signal Service. The price was fixed at $25,000, but a bonus of $2500 was to be paid for every mile above forty miles an hour made by the machine on its trial trip. That bonus looked big to the Wrights, but it cost the cause of aviation many times its face value in the congressional disfavour it caused. Aviation was then in its infancy in the United States. Every man in Congress wanted to see the flights. But Fort Myer, whose parade was to be the testing ground, was fully fourteen miles from the Capitol, and reached only most inconveniently from Washington by trolley, or most expensively by carriage or automobile. Day after day members of the House and Senate made the long journey across the Potomac. Time and again they journeyed back without even a sight of the flyer in the hangar. One after another little flaws discovered in the machine led the aviators to postpone their flight. Investigating statesmen who thought that their position justified them in seeking special privileges were brusquely turned away by the military guard. The dusk of many a summer's night saw thousands of disappointed sightseers tramping the

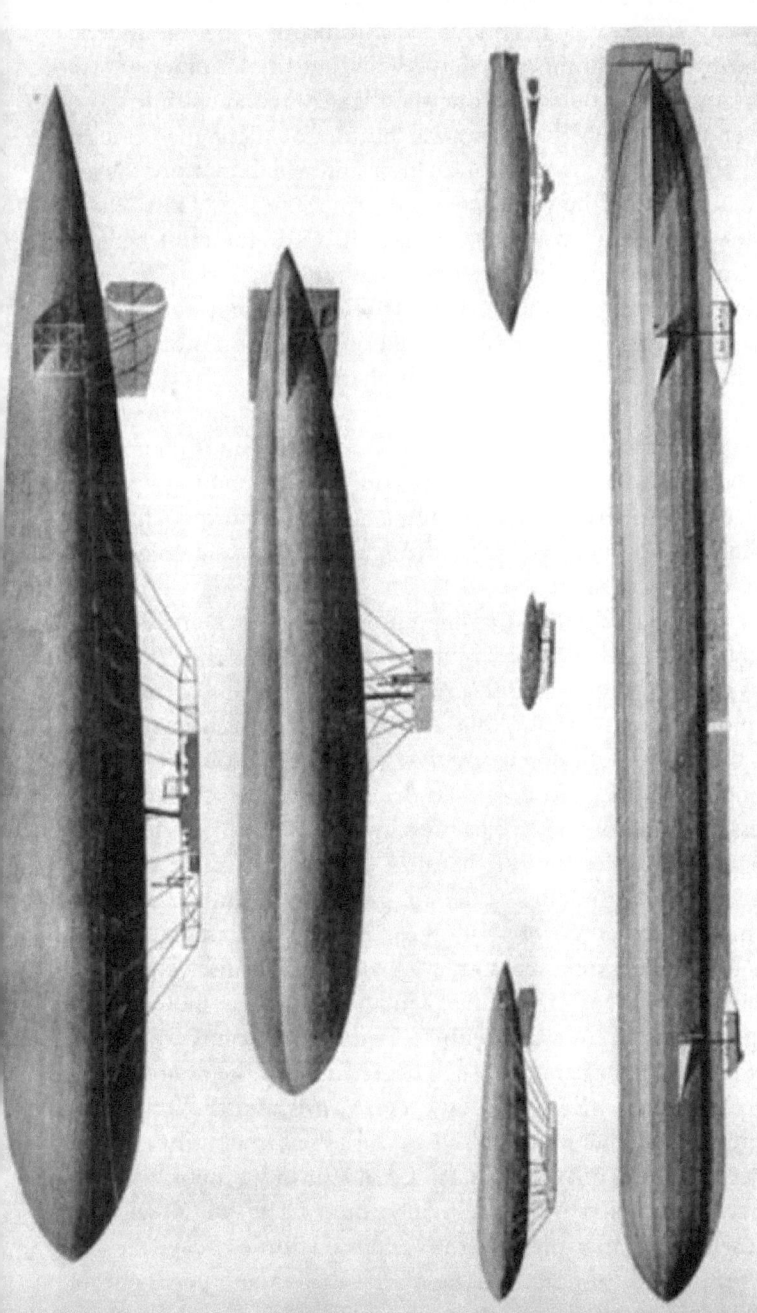

THE COMPARATIVE STRENGTH OF BELLIGERENTS IN DIRIGIBLES AT THE OPENING OF THE WAR.

France must be credited with at least eighteen airships of various types—England had only seven—Russia had probably not more than three airships available—Belgium had one airship Austria had not less than three, not more than five airships available—Germany had twenty three airships of the rigid, semi-rigid, and non-rigid type.

long road back to Washington. The climax came when on a clear but breezy day Wilbur Wright announced that the machine was in perfect condition and could meet its tests readily, but that in order to win a bigger bonus, he would postpone the flight for a day with less wind. All over Washington the threat was heard that night that Congress would vote no more money for aviation, and whether or not the incident was the cause, the sequence was that the American Congress was, until the menace of war with Germany in 1916, the most niggardly of all legislative bodies in its treatment of the flying corps. When the Wrights did finally fly they made a triumphant flight before twelve thousand spectators. The test involved crossing the Potomac, going down its north side to Alexandria, and then back to Fort Myer. Ringing cheers and the crashing strains of the military band greeted the return of the aviator, but oblivious to the enthusiasm Wilbur Wright stood beside his machine with pencil and pad computing his bonus. It figured up to five thousand dollars, and the reporters chronicled that the Wrights knew well the difference between solid coin and the bubble of reputation.

But this seemingly cold indifference to fame and single-minded concentration on the business of flying on the part of the Wrights was in fact of the utmost value to aviation as an art and a science. They were pioneers and successful ones. Their example was heeded by others in the business. In every way they sought to discourage that wild reaching after public favour and notoriety that led aviators to attempt reckless feats, and often sacrifice their lives in a foolish effort to astonish an audience. No one ever heard of either of the Wright brothers "looping-the-loop," doing a "demon glide," or in any other fashion reducing the profession of aviation to the level of a circus. In a time when brave and skilful aviators, with a mistaken idea of the ethics of their calling, were appealing to sensation lovers by the practice of dare-devil feats, the Wrights with admirable common sense and dignity stood sturdily against any such degradation of the aviator's art. In this position they were joined by Glenn Curtis, and the influence of the three was beginning to be shown in the reduced number of lives sacrificed in these follies when the Great War broke upon the world and gave to aviation its greatest opportunity. The world will hope nevertheless that after that war shall end the effort to adapt the aeroplane to the ends of peace will be no less earnest and persistent than have been the methods by which it has been made a most serviceable auxiliary of war.

WRIGHT GLIDER

In July, 1915, *Collier's Weekly* published an interview with Orville Wright in which that man, ordinarily of few words, set up some interesting theories upon the future of aeroplanes.

"The greatest use of the aeroplane to date," said Mr. Wright, "has been as a tremendously big factor of modern warfare. But—

"The greatest use of the aeroplane eventually will be to prevent war.

"Some day there will be neither war nor rumours of war, and the reason may be flying machines.

"It sounds paradoxical. We are building aeroplanes to use in time of war, and will continue to build them for war. We think of war and we think of aeroplanes. Later on, perhaps, we shall think of aeroplanes in connection with the wisdom of keeping out of war.

"The aeroplane will prevent war by making it too expensive, too slow, too difficult, too long drawn out—in brief, by making the cost prohibitive.

"Did you ever stop to think," inquires Wright, "that there is a very definite reason why the present war in Europe has dragged along for a year with neither side gaining much advantage over the other? The reason as I figure it out is aeroplanes. In consequence of the scouting work done by the flying machines each side knows exactly what the opposing forces are doing.

"There is little chance for one army to take another by surprise. Napoleon won his wars by massing his troops at unexpected places. The aeroplane has made that impossible. It has equalized information. Each side has such complete knowledge of the other's movements that both sides are obliged to crawl into trenches and fight by means of slow, tedious routine, rather than by quick, spectacular dashes.

"My impression is that before the present war started the army experts expected it to be a matter of a few weeks, or at the most, a few months. To-day it looks as if it might run into years before one side can dictate terms. Now, a nation that may be willing to undertake a war lasting a few months may well hesitate about engaging in one that will occupy years. The daily cost of a great war is of course stupendous. When this cost runs on for years the total is likely to be so great that the side which wins nevertheless loses. War will become prohibitively expensive. The scouting work in flying machines will be the predominating factor, as it seems to me, in bringing this about. I like to think so anyhow."

"What, in your opinion, has the present war demonstrated regarding the relative advantages of aeroplanes and Zeppelin airships?" the inventor was asked.

"The aeroplane seems to have been of the more practical use," replied Wright. "In the first place, dirigible airships of the Zeppelin type are so expensive to build, costing somewhere around a half million dollars each, that it is distinctly disadvantageous to the nation operating them to have one destroyed. But what is more important is the fact that the Zeppelin is so large that it furnishes an excellent target, unless it sails considerably higher than is comparatively safe for an aeroplane. And when the Zeppelin is at a safe height it is too far above the ground for your scout to make accurate observations. Similarly, when the Zeppelin is used for dropping bombs, it must be too high for the bomb thrower to show much accuracy."

"You think that the use of flying machines for scouting purposes will be of considerably more importance than their use as a means of attack?" was another question.

"That has been decidedly true so far," replied Wright. "About all that has been accomplished by either side from bomb dropping has been to kill a few non-combatants and that will have no bearing on the result of the war.

AT A FRENCH AEROPLANE BASE

"English newspapers have long talked of the danger of Zeppelin attacks or aeroplane attacks, but it was all for a purpose, because they did not believe the country was sufficiently prepared for war and sought to arouse the people and the War Department to action by means of the airship bogy. (Later history showed Mr. Wright sadly in error on this point.)

"Aside from the use of the machines for war purposes the war will give a great boost to aviation generally. It has led more men to learn to fly, and with a higher degree of skill than ever before. It has awakened people to aviation possibilities.

"Just like the automobile, it will become more and more fool-proof, easier to handle and safer. There is no reason why it should not take the place of special trains where there is urgent need of great speed.

"The aeroplane has never really come into its own as a sporting proposition. Of late years the tendency has been to develop a high rate of speed rather than to build machines that may be operated safely at a comparatively low speed. You see, a machine adapted to make from seventy to one hundred miles an hour cannot run at all except at a pretty rapid clip, and this means difficulty in getting down. One must have a good, smooth piece of ground to land on and plenty of it. When we get an aeroplane that will fly along at twenty miles an hour, one can land almost any place,—on a roof, if necessary,—and then people will begin to take an interest in owning an aeroplane for the enjoyment of flying."

"Is it true that you and your brother had a compact not to fly together?"

"Yes, we felt that until the records of our work could be made complete it was a wise precaution not to take a chance on both of us getting killed at the same time. We never flew together but once. From 1900 to 1908 the total time in the air for both Wilbur and myself, all put together, was only about four hours."

Mr. Wright's statement of the brevity of the time spent in actual flying in order to learn the art will astonish many people. Few novices would be so rash as to undertake to steer an automobile alone after only four hours' practice, and despite the fact that the aviator always has plenty of space to himself the aeroplane can hardly yet be regarded as simple a machine to handle as the automobile. Nev-

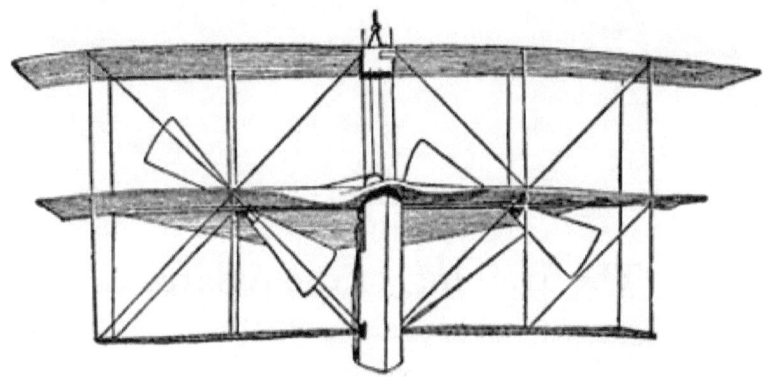

STRINGFELLOW'S AEROPLANE

ertheless the ease with which the method of its actual manipulation is acquired is surprising. More work is done in the classroom and on the ground to make the fighting pilot than in the air. As we have traced the development of both dirigible and aeroplane from the first nascent germ of their creation to the point at which they were sufficiently developed to play a large part in the greatest of all wars, let us now consider how hosts of young men, boys in truth, were trained to fly like eagles and to give battle in mid-air to foes no less well trained and desperate than they.

CHAPTER 6

The Training of the Aviator

The Great War, opening in Europe in 1914 and before its end involving practically the whole world, including our own nation, has had more to do with the rapid development of aircraft, both dirigible balloons and aeroplanes, than any other agency up to the present time. It tested widely and discarded all but the most efficient. It established the relative value of the dirigible and the aeroplane, so relegating the former to the rear that it is said that the death of Count Zeppelin, March 8, 1917, was in a measure due to his chagrin and disappointment. It stimulated at once the inventiveness of the constructors and the skill and daring of the pilots. When it opened there were a few thousand machines and trained pilots in all the armies of Europe. Before the war had been in progress three years there were more flying men over the battlefields of the three continents, Europe, Asia, and Africa, than there were at that time soldiers of all classes enlisted in the regular army of the United States. Before that war the three arms of the armed service had been infantry, artillery, and cavalry. The experience of war added a new arm—the aviation corps—and there is to-day some doubt whether in importance it should not be ranked above the cavalry.

When war was declared none of the belligerent nations had its aerial fleet properly organized, nor was the aviation department in any of them equal in preparedness to the rest of the army. The two great antagonists did not differ greatly in the strength of their flying forces. Germany possessed about 1000 aeroplanes, exclusive of about 450 in private hands, of all which it is estimated about 700 were ready for immediate service. Fourteen Zeppelins were in commission, and other large dirigibles of different types brought the number of the craft of this sort available up to forty.

France was stronger in aeroplanes but weaker in dirigibles. Of the

AMERICA—BUILT TO CROSS THE ATLANTIC OCEAN

former she had about 1500; of the latter not more than twenty-five. The land was swept for planes in the hands of private owners and, as the French people had from the first taken a lively interest in aviation, more than 500 were thus obtained. The French furthermore at the very outset imperilled their immediate strength in the air for the sake of the future by adopting four or five machines as army types and throwing out all of other makes. More than 550 machines were thus discarded, and their services lost during the first weeks of the war. The reason for this action was the determination of the French to equip their aviation corps with standardized machines of a few types only. Thus interchangeable parts could always be kept in readiness in case of an emergency, and the aviation corps was obliged to familiarize itself with the workings of only a few machines. The objection to the system is the fact that it practically stopped all development of any machines in France except the favoured few. Moreover it threw out of the service at a stroke, or remanded for further instruction, not less than four hundred pilots who had been trained on the rejected machines. The order was received with great public dissatisfaction, and for a time threatened serious trouble in the Chamber of Deputies where criticisms of the direction of the flying service even menaced the continuance of the ministry in power.

At the outset of the war Great Britain lagged far behind the other chief belligerents in the extent of her preparations for war in the air. As has been pointed out the people of that nation had never taken the general interest in aviation which was manifested in France, and there was no persistent Count von Zeppelin to stir government and citizens into action. The situation was rather anomalous. Protected from invasion by its ring of surrounding waters, England had long concentrated its defensive efforts upon its navy. But while the danger of invasion by the air was second only to that by sea the British contemplated with indifference the feverish building of Zeppelins by Germany, and the multiplication of aircraft of every sort in all the nations of the continent. The manufacture of aircraft was left to private builders, and not until the war was well under way did the government undertake its systematic supervision. The Royal Aerial Factory, then established, became the chief manufacturer of machines for army and navy use, and acted also as the agent for the inspection and testing of machines built by private firms. Control of the Royal Flying Corps is vested in the Admiralty, the government holding that the strategy of airships was distinctly naval.

WRIGHT AEROPLANE IN FLIGHT

In the use of seaplanes the British were early far in the lead of other nations, as we shall see in a later chapter. And in the prompt and efficient employment of such aircraft as she possessed at the opening of the war she far outclassed Germany which in point of numbers was her superior. At that moment Great Britain possessed about five hundred machines, of which two hundred were seaplanes, and fifteen dirigibles. Despite this puny force, however, British aviators flew across the channel in such numbers to the headquarters in France that when the Expeditionary Army arrived on the scene it found ready to its hand a scouting force vastly superior to anything the Germans could put in the air. It is no exaggeration to say that the Royal Flying Corps saved Sir John French's army in his long and gallant fight against the overwhelming numbers of the foe.

Russia before the war had hidden her aeronautic activities behind the dreary curtain of miles of steppe and marsh that shut her off from the watchfulness of Western Europe. Professional aviators, indeed, had gone thither to make exhibition flights for enormous purses and had brought back word of huge aeroplanes in course of construction and an eager public interest in the subject of flying. But the secrecy which all the governments so soon to be plunged in war sought to throw about their production of aircraft was especially easy for Russia in her isolation. When the storm burst her air fleet was not less than eight hundred aeroplanes, and at least twenty-five dirigibles.

A competent authority estimates that at the outbreak of the war the various Powers possessed a total of 4980 aircraft of all sorts. This sounds like a colossal fleet, but by 1917 it was probably multiplied more than tenfold. Of the increase of aircraft we can judge only by guesswork. The belligerents keep their output an inviolable secret. It was known that many factories with a capacity of from thirty to fifty 'planes a week were working in the chief belligerent lands, that the United States was shipping aircraft in parts to avoid violation of neutrality laws before their entrance upon the war, and that American capital operated factories in Canada whence the completed craft could be shipped regardless of such laws. How great was the loss to be offset against this new construction is a subject on which no authoritative figures are available.

It was estimated early in the war that the life of an aeroplane in active service seldom exceeded three weeks. In passing it may be mentioned that by some misapprehension on the part of the public, this estimate of the duration of a machine was thought to cover also the average life of the aviators in service. Happily this was far from true. The mortality among the machines was not altogether due to wounds sustained in combat, but largely to general wear and tear, rough usage, and constant service. The slightest sign of weakness in a machine led to its instant condemnation and destruction, for if it should develop in mid-air into a serious fault it might cost the life of the aviator and even a serious disaster to the army which he was serving. As the war went on the period of service of a machine became even briefer, for with the growing demand for faster and more quickly controllable machines everything was sacrificed to lightness and speed. The factor of safety which early in the war was six to eight was reduced to three and a half, and instances were known in all services of machines simply collapsing and going to pieces under their own weight without wound or shock.

About the extent to which the belligerent governments developed their air forces after the outbreak of war there was during the continuance of that conflict great reticence maintained by all of them. At the outset there was little employment of the flyers except on scouting reconnaissance work, or in directing artillery fire. The raids of Zeppelins upon England, of seaplanes on Kiel and Cuxhaven, of aeroplanes on Friedrichshaven, Essen, and Venice came later. It has been noted by military authorities that, while Germany was provided at first with the largest aviation force of all the belligerents, she either underestimated its value at the outset, or did not know how to employ it, for she blundered into and through Belgium using her traditional Uhlans for scouts, to the virtual exclusion of airmen. The effectiveness of the Belgian fight for delay is ascribed largely to the intelligent and effective use its strategists made of the few aircraft they possessed.

Wellington was wont to say that the thing he yearned for most in battle was to "see the other side of that hill."

Napoleon wrote:

Nothing is more contradictory, nothing more bewildering than the multitude of reports of spies, or of officers sent out to reconnoitre. Some locate army corps where they have seen only detachments; others see only detachments where they ought to have seen army corps.

So the two great protagonists of the opening years of the nineteenth century deplored their military blindness. In the opening years of the twentieth it was healed. All that Wellington strove to see, all that

The Lafayette Escadrille—first Americans to fly in France. (Lufbery on left, Thaw on right.)

the cavalry failed to find for Napoleon is to-day brought to headquarters by airmen, neatly set forth in maps, supported by photographs of the enemy's positions taken from the sky.

Before describing the exploits of the airmen in actual campaign let us consider some account of how they were trained for their arduous and novel duties.

To the non-professional an amazing thing about the employment of aircraft in war has been the rapidity with which pilots are trained. The average layman would think that to learn the art of manoeuvring an aeroplane with such swiftness as to evade the attacks of an enemy, and to detect precisely the proper moment and method of attacking him in turn, would require long and arduous practice in the air. But as we have seen in earlier chapters, inventors like the Wrights, Bleriot, and Farman learned to fly with but a few hours spent in the air, with flights lasting less than ten minutes each. So too the army aviators spent but little time aloft, though their course of instruction covered in all a period of about four months.

Some account of the method of instruction as reported by several out of the hundred or more American boys who went to fly for France may be interesting.

As a rule the aviators were from twenty to twenty-five years of age. "Below twenty boys are too rash; above twenty-five they are too prudent," said a sententious French aviator. A slight knowledge of motors such as would be obtained from familiarity with automobiles was a marked advantage at the start, for the first task of the novice was to make himself familiar with every type of aeroplane engine. The army pilot in all the armies was the aristocrat of the service. Mechanics kept his motor in shape, and helpers housed, cleaned, and brought forth his machine for action. But while all but the actual piloting and fighting was spared him, there was always the possibility of his making an untimely landing back of the enemy's lines with an engine that would not work. To prepare for such an emergency he was taught all the intricacies of motor construction, so that he might speedily correct any minor fault.

In our army, and indeed in all others, applicants for appointment to the aviation corps were subjected to scientific tests of their nerves, and their mental and physical alertness. How they would react to the sudden explosion of a shell near their ears, how long it took the candidate to respond to a sudden call for action, how swiftly he reacted to a sensation of touch were all tested and measured by delicate electric apparatus. A standard was fixed, failing to attain which, the applicant was rejected. The practical effect might be to determine how long after suddenly discovering a masked machine gun a given candidate would take before taking the action necessary to avoid its fire. Or how quickly would he pull the lever necessary to guard against a sudden gust of wind. To the layman it would appear that problems of this sort could only be solved in the presence of the actual attack, but science, which enables artillerists to destroy a little village beyond the hills which they never see, was able to devise instruments to answer these questions in the quiet of the laboratory.

One of the best known flying schools of the French army was at Pau, where on broad level plains were, in 1917, four separate camps for aviators, each with its group of hangars for the machines, its repair shops, and with a tall wireless tower upstanding in the midst for the daily war news from Paris. On these plains the Wright Brothers had made some of their earliest French flights. A little red barn which they had made their workshop was still standing there when war suddenly turned the spot into a flying school often with as many as five thousand pupils in attendance. "To-day that little red barn," writes Carroll Dana Winslow, one of the Americans who went to fly for France, "stands as a monument to American stupidity, for when we allowed the Wrights to go abroad to perfect their ideas instead of aiding them to carry on their work at home we lost a golden opportunity. Now the United States which gave to the world the first practical aeroplane is the least advanced in this all-important science."

Arrived at the school the tyro studies the fundamentals of flying in the classroom and on the field for two months before he is allowed to go up—to receive as they express it, his *baptême de l'air*. He picks motors to pieces, and puts them together, he learns the principles of aeroplane construction, and can discourse on such topics as the angle of attack of the cellule, the incidence of the wings, and the carrying power of the tail-plane. More than any other science aviation has a vocabulary of its own, and a peculiarly cosmopolitan one drawn from all tongues, but with the French predominating. America gave the aeroplane to France, but France has given the science its terminology.

The maps of the battlefields of this war are the marvels of military science. Made from the air they show every road and watercourse, every ditch and gully, every patch of woodland, every farmhouse, church, or stonewall. Much of the early work of the aviator is in learning to make such maps, both by sketches and by the employment of the camera. It is no easy task. From an aeroplane one thousand feet up the earth seems to be all a dead level. Slight hills, gentle elevations, offer no contrast to the general plain. A road is not easy to tell from a trench. All these things the aviator must first learn to see with accuracy, and then to depict on his map with precision. He must learn furthermore to read the maps of his fellows—a task presupposing some knowledge of how they had been made. He must learn to fly by a map, to recognize objects by the technical signs upon it, to estimate his drift before the wind because of which the machine moves sidewise *en crabe*—or like a crab as the French phrase it.

His first flight the novice makes in a machine especially fitted for instruction. The levers are fitted with double handles so that both learner and tutor may hold them at once. If the greenhorn pushes when he should pull the veteran's grip is hard on the handle to correct the error before it can cost two lives—for in the air there is little time to experiment. Either set of controls will steer the machine. The pupil grasps his levers, and puts his feet on the pedals. At first the instructor will do the steering, the pupil following with hands and feet as the motions made by the instructor are communicated to him by the moving levers. For a time the two work together. Then as the instructor senses that the student himself is doing the right thing he gradually lessens his own activity, until after a few days' practice the student finds that he is flying with a passenger and directing the machine himself. In France, at any rate, they teach in brief lessons. Each flight for instruction is limited to about five minutes. At first the student operates in a "penguin"—a machine which will run swiftly along the ground but cannot rise. It is no easy trick at first, to control the "penguin" and keep its course direct. Then he will try the "jumps" in a machine that leaps into the air and descends automatically after a twenty to forty yards' flight. As Darius Green expressed it so long ago, the trouble about flying comes when you want to alight. That holds as true to-day with the most perfect aeroplanes, as in boyhood days when one jumped from the barn in perfect confidence that the family umbrella would serve as a parachute. To alight with an aeroplane the pilot—supposing his descent to be voluntary and not compelled by accident or otherwise—surveys the country about him for a level field, big and clear enough for the machine to run off its momentum in a run of perhaps two hundred yards on its wheels. Then he gets up a good rate of speed, points the nose of the 'plane down at a sharp angle to the ground, cuts off the engine, and glides. The angle of the fall must be great enough for the force of gravity to keep up the speed. There is a minimum speed at which an aeroplane will remain subject to control. Loss of speed—*perte de vitesse*, as the French call it—is the aviator's most common peril in landing. If it occurs after his engine is cut off and he has not the time to start it again, the machine tilts and slides down sideways. If it occurs higher up a *vrille* is the probable result. In this the plane plunges toward the ground spinning round and round with the corner of one wing as a pivot. In either case a serious accident is almost inevitable.

In fact the land is almost as dangerous to the navigator of the air as it is to him of the sea. To make good landings is an art only perfected by constant practice. To shut off the engine at precisely the right moment, to choose an angle of descent that will secure the greatest speed and at the same moment bring you to your landing place, to change at the most favourable time from this angle to one that will bring you to the ground at the most gentle of obtuse angles, and to let your machine, weighing perhaps a ton, drop as lightly as a bird and run along the earth for several hundred feet before coming to a full stop, are all features of making a landing which the aviator has to master.

In full air there are but few perils to encounter. All airmen unite in declaring that even to the novice in an aeroplane there is none of that sense of dizziness or vertigo which so many people experience in looking down from high places. The flyer has no sense of motion. A speed of forty miles an hour and of one hundred miles are the same to him. As he looks down the earth seems to be slipping away from him, and moving by, tailwards, like an old-fashioned panorama being unwound.

Everything about the control of an aeroplane has to be learned mechanically. Once learned the aviator applies his knowledge intuitively. He "senses" the position and progress of the craft by the feel of the controls, as the man at the yacht's tiller tells mysteriously how she is responding to the breeze by "the feel." Even before the 'plane responds to some sudden gust of wind, or drops into a hole in the air, the trained aviator will foresee precisely what is about to happen. He reads it in some little thrill of his lever, a quiver in the frame, as the trained boxer reads in his antagonist's eyes the sort of blow that is coming. This instinctive control of his machine is absolutely essential for the fighting pilot who must keep his eyes on the movements of his enemy, watch out for possible aircraft guns below, and all the time be striving to get an advantageous position whence he can turn his machine gun loose. A row of gauges, dials, a compass, and a map on the frame of the car in which he sits will engage his attention in any moments of leisure. It is needless to remark that the successful pilot must have a quick eye and steady nerves.

Nerve and rapidity of thought save the aviator in many a ticklish position. It is perhaps a tribute to the growing perfection of the aeroplanes that in certain moments of peril the machine is best left wholly to itself. Its stability is such that if freed from control it will often right itself and glide safely to earth. This not infrequently occurs in the moment of the dreaded *perte de vitesse*, to which reference has been made. In his book, *With the French Flying Corps*, Mr. Carroll Dana Winslow, a daring American aviator, tells of two such experiences, the one under his observation, the other happening to himself:

> The modern aeroplane is naturally so stable that if not interfered with it will always attempt to right itself before the dreaded *vrille* occurs, and fall *en feuille morte*. Like a leaf dropping in an autumn breeze is what this means, and no other words explain the meaning better.
>
> A curious instance of this happened one day as I was watching the flights and waiting for my turn. I was particularly interested in a machine that had just risen from the Grande Piste. It was acting very peculiarly. Suddenly its motor was heard to stop. Instead of diving it commenced to wobble, indicating a *perte de vitesse*. It slipped off on the wing and then dove. I watched it intently, expecting it to turn into the dreaded spiral. Instead it began to climb. Then it went off on the wing, righted itself, again slipped off on the wing, volplaned, and went off once more. This extraordinary performance was repeated several times, while each time the machine approached nearer and nearer to the ground. I thought that the pilot would surely be killed. Luck was with him, however, for his slip ceased just as he made contact with the ground and he settled in a neighbouring field. It was a very bumpy landing but the aeroplane was undamaged.
>
> The officers rushed to the spot to find out what was the matter. They found the pilot unconscious, but otherwise unhurt. Later in the hospital he explained that the altitude had affected his heart and that he had fainted. As he felt himself going

he remembered his instructions and relinquished the controls, at the same time stopping his motor. His presence of mind and his luck had saved his life—his luck I say, for had the machine not righted itself at the moment of touching the ground it would have been inevitably wrecked.

The spectacle, though terrifying, proved valuable as an education to young Winslow who a few days later was ordered to a test of ascension of two thousand feet. This is his story:

I had a narrow escape. I had received orders to make a flight during a snow-storm. I rose to the prescribed height and then prepared to make my descent. A whirling squall caught me in the act of making a spiral. I felt the tail of my machine go down and the nose point up. I had a classical *perte de vitesse*. I looked out and saw that I was less than eight hundred feet above the ground and approaching it at an alarming rate of speed. I had already shut off the motor for the spiral, and turning it on, I knew, would not help me in the least. Suddenly I remembered the pilot who fainted. I let go of everything, and with a sickening feeling I looked down at the up-rushing ground. At that instant I felt the machine give a lurch and right itself. I grabbed the controls, turned on the motor, and resumed my line of flight only two hundred feet in the air. All this happened in a few seconds, but my helplessness seemed to have lasted for hours. I had had a very close call—not as close as the man who fainted, but sufficiently so for me.

We have said that the process of training a flyer is remarkably expe-

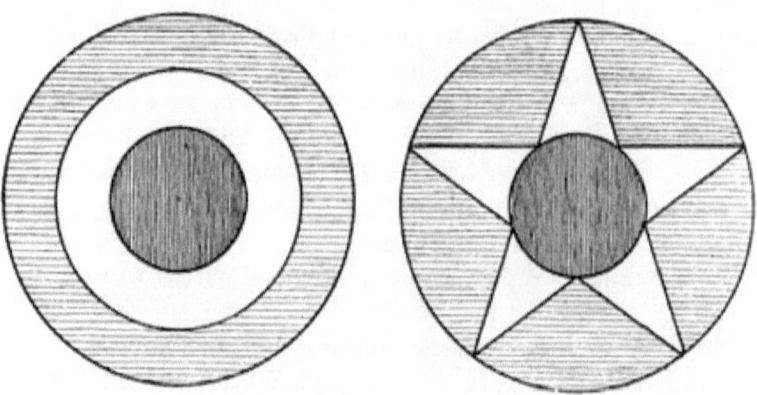

DISTINGUISHING MARKS OF AMERICAN PLANES

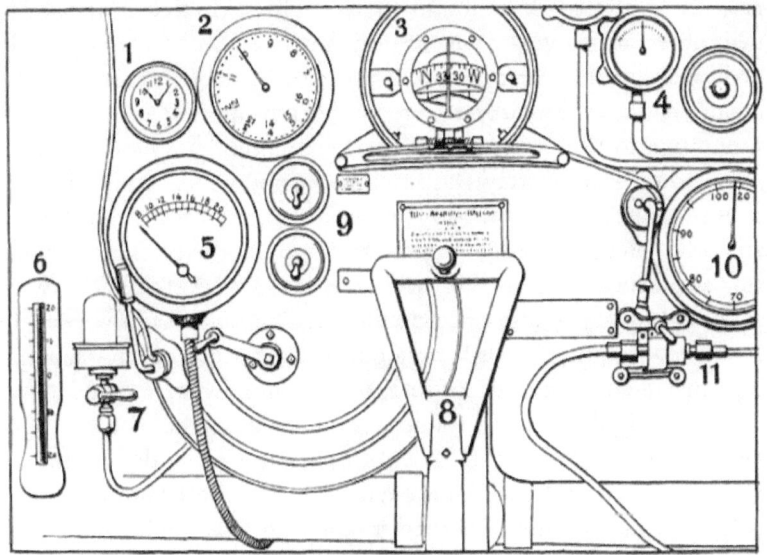

WHAT AN AVIATOR MUST WATCH.
1 *Watch* 2 *Altimeter-registering height* 3 *Compass* 4 *Pressure gauges for two gasoline tanks* 5 *Dial registering engine revolutions* 6 *Inclinometer, registering level fore and aft* 7 *Oil pulsator* 8 *Control stick, with thumb switch* 9 *Switches, two magnetos* 10 *Air speed indicator* 11 *Gasoline supply pipe*

ditious. So far as the fundamentals of his profession are concerned it is. But his education in fact never ends. In the mere matter of reconnaissance, for example, experience is everything. One might imagine that ten thousand men marching on a road would look alike in numbers whatever the nationality. Not so. To the untrained eye five thousand or six thousand French troops will look as numerous as ten thousand British or Germans. Why? Because the French march in much more extended order. Into their democratic military methods the precision and mechanical exactitude of German drill do not enter. With the same number of troops they will extend further along the road by at least a third than would a detachment of either of the other armies.

And again. Great skill has been developed in the course of the war in the art of concealing positions and particularly in disguising cannon. The art has given a new word to the world—*camouflage*. Correspondents have repeatedly told of their amazement in suddenly coming across a battery of 75's, or a great siege gun so cunningly hidden in the edge of a thicket they would be almost upon it before detecting it. From an aeroplane 2500 feet or more in the air it requires sharp eyes to penetrate artillery disguises. A French *poilu* in a little book of reminiscences tells with glee how a German observation aviator deceived his batteries. A considerable body of French troops being halted in an open field, out of sight of the enemy batteries, found the glare of the sun oppressive, and having some time to wait threw down their equipment and betook themselves to the cool shadows of a neighbouring wood. Along came an enemy aviator. From his lofty height the haversacks, blanket-rolls, and other pieces of dark equipment lying upon the grass looked like a body of troops resting. After sailing over and around the field twice as though to make assurance doubly sure he sailed swiftly away. In a very few minutes shells from a concealed battery began dropping into that field at the rate of several a minute. Every foot of it was torn up, and the French soldiers from their retreat in the woods saw their equipment being blown to pieces in every direction. The spectacle was harrowing, but the reflection that the aviator undoubtedly thought that he had turned his guns on a field full of men was cheering to them in their safety.

An art which the fighting aviator must master early in his career is that of high diving. Many of us have seen a hawk, soaring high in air, suddenly fold his pinions and drop like a plummet full on the back of some luckless pigeon flapping along ungainly scores of feet below, or a fishhawk drop like a meteor from the sky with a resounding splash upon the bosom of some placid stream and rise again carrying a flapping fish to his eyrie in the distant pines. The hunting methods of the hawk are the fighting methods of the airman. But his dives exceed in height and daring anything known to the feathered warriors of the air.

Boelke, most famous of all the German airmen—or for that matter of all aerial fighters of his day—who in 1917 held the record for the number of enemy flyers brought down, was famed for his savage dives. He would fly at a great height, fifteen thousand or more feet, thus assuring himself that there was no enemy above him. When he sighted his prey he would make an absolutely vertical nose dive, dropping at the rate of 150 miles an hour or more and spattering shots from his machine gun as he fell. Six hundred shots a minute and the sight of this charging demon were enough to test the nerve of any threatened aviator. In some fashion Boelke was enabled to give a slight spiral form to his dive so that his victim was enveloped in a ring of bullets that blocked his retreat whichever way he might turn for safety.

Personality in fighting counted much for success. Boelke's method, its audacity and fierceness, placed him first in the list of airmen with killing records. Captain Immelman, also a German, who rolled up a score of thirty enemies put out of action before he himself was slain, followed entirely different tactics. His battle manoeuvre savoured much of the circus, including as it did complete loop-the-loop. For instead of approaching his adversary from the side, or as would be said in the sea navy, on the beam, he followed squarely behind him. His study was to get the nose of his machine almost on the tail of the aircraft he was pursuing. This gave him, to begin with, what used to be called in the navy a raking position, for his shots would rake the whole body of the enemy aeroplane from tail to nose with a fair chance of hitting either the fuel tank, the engine, or the pilot. Failing to secure the position he most coveted, this daring German would surrender it with apparent unconcern to the enemy who usually fell into the trap. For just as the foeman's machine came up to the tail of Immelman's craft the latter would suddenly turn his nose straight to earth, drop like a stone, execute a backward loop, and come up behind his surprised adversary who thus found the tables suddenly turned.

These two German aviators long held the record for execution done in single combat. Boelke was killed before the air duel vanished to be replaced by the battle of scores of planes high in air. Immelman survived longer, but with the incoming of the pitched battle his personal prowess counted for less and his fame waned.

In July, 1917, arrangements were complete in the United States for the immediate training in the fundamentals of aviation of ten thousand young Americans. The expectation was that long before the end of the year facilities would be provided for the training of many more. Both France and Great Britain sent over squads of their best aviators, some of them so incapacitated from wounds as to be disqualified for further fighting, but still vigorous enough for the work of an instructor. The aerial service took hold upon the imagination and the patriotism of young America as did no other. The flock of volunteers was far beyond the capacity of the government to care for, and many drifted over into private aviation schools which were established in great numbers. The need for the young students was admittedly great. More and more the impression had grown in both Great Britain and France that the aeroplane was to be the final arbiter in the war. It was hailed at once as the most dangerous enemy of the submarine and the most efficient ally of troops in the field. No number seemed too great for the needs of the entente allies, and their eagerness to increase their flying force was strengthened by the knowledge of the fact that Germany was building feverishly in order that its fleet in the air might not be eclipsed.

Perhaps the best description of an idealized aviator was given by Lieutenant Lufbery, of the Lafayette Escadrille, who came to the United States to assist in training the new corps of American flying men. Lufbery himself was a most successful air fighter—an "ace" several times over. Though French by lineage, he was an American citizen and had been a soldier in the United States Army. In October of 1917 his record was thirteen Boches brought down within the allied lines. In the allied air service one gets no credit for the defeated enemy plane if it falls within the enemy lines.

While young Americans were being drilled into shape for service in the flying corps, Lufbery gave this outline of the type of men the service would demand:

> It will take the cream of the American youth between the ages of eighteen and twenty-six to man America's thousands of aeroplanes, and the double cream of youth to qualify as chasers in the Republic's new aerial army.
>
> Intensive and scientific training must be given this cream of youth upon which America's welfare in the war must rest. Experience has shown that for best results the fighting aviator should be not over twenty-six years old or under eighteen. The youth under eighteen has shown himself to be bold, but he lacks judgment. Men over twenty-six are too cautious.
>
> The best air fighters, especially a man handling a chaser, must be of perfect physique. He must have the coolest nerve and be of a temperament that longs for a fight. He must have a sense of absolute duty and fearlessness, the keenest sense of action, and perfect sight to gain the absolute "feel" of his machine.
>
> He must be entirely familiar with aerial acrobatics. The latter frequently means life or death.
>
> Fighting twenty-two thousand feet in the air produces a heavy strain on the heart. It is vital therefore that this organ show not the slightest evidence of weakness. Such weakness would decrease the aviator's fighting efficiency.
>
> The American boys who come over to France for this work will be subject to rapid and frequent variations in altitude. It is

a common occurrence to dive vertically from six thousand to ten thousand feet with the motor pulling hard.

Sharpness of vision is imperative. Otherwise the enemy may escape or the aviator himself will be surprised or mistake a friendly machine for a hostile craft. The differences are often merely insignificant colours and details.

America's aviators must be men who will be absolute masters of themselves under fire, thinking out their attacks as their fight progresses.

Experience has shown that the chaser men should weigh under 180 pounds. Americans from the ranks of sport, youth who have played baseball, polo, football, or have shot and participated in other sports will make the best fighting aviators.

CHAPTER 7

Some Methods of the War in the Air

The fighting tactics of the airmen with the various armies were developed as the war ran its course. As happens so often in the utilization of a new device, either of war or peace, the manner of its use was by no means what was expected at the outset. For the first year of the war the activities of the airmen fell far short of realizing Tennyson's conception of

The nations' airy navies grappling in the central blue.

The grappling was only incidental. The flyers seemed destined to be scouts and rangefinders, rather than fighters. Such pitched combats as there were took rather the form of duels, conducted with something of the formality of the days of chivalry. The aviator intent upon a fight would take his machine over the enemy's line and in various ways convey a challenge to a rival—often a hostile aviator of fame for his daring and skill in combat. If the duel was to the death it would be watched usually from the ground by the comrades of the two duellists, and if the one who fell left his body in the enemy's lines, the victor would gather up his identification disk and other personal belongings and drop them the next day in the camp of the dead man's comrades with a note of polite regret.

It was all very daring and chivalric, but it was not war according to twentieth century standards and was not long continued.

When at first the aviators of one side flew over the enemy's territory diligently mapping out his trenches, observing the movements of his troops, or indicating, by dropping bunches of tinsel for the sun to shine upon or breaking smoke bombs, the position of his hidden battery, the foe thus menaced sought to drive them away with anti-aircraft guns. These proved to be ineffective and it may be said here that throughout the war the swift aeroplanes proved themselves more

A Caproni triplane

than a match for the best anti-aircraft artillery that had been devised. They could complete their reconnaissances or give their signals at a height out of range of these guns, or at least so great that the chances of their being hit were but slight. It was amazing the manner in which an aeroplane could navigate a stretch of air full of bursting shrapnel and yet escape serious injury. The mere puncture, even the repeated puncture, of the wings did no damage. Only lucky shots that might pierce the fuel tank, hit the engine, touch an aileron or an important stay or strut, could affect the machine, while in due course of time a light armour on the bottom of the fuselage or body of the machine in which the pilot sat, protected the operator to some degree. Other considerations, however, finally led to the rejection of armour.

Accordingly it soon became the custom of the commanders who saw their works being spied out by an enemy soaring above to send up one or more aircraft to challenge the invader and drive him away. This led to the second step in the development in aerial strategy. It was perfectly evident that a man could not observe critically a position and draw maps of it, or seek out the hiding place of massed batteries and indicate them to his own artillerists, and at the same time protect himself from assaults. Accordingly the flying corps of every army gradually became differentiated into observation machines and fighting machines—or *avions de réglage, avions de bombardement,* and *avions de chasse,* as the French call them. In their order these titles were applied to heavy slow-moving machines used for taking photographs and directing artillery fire, more heavily armed machines of greater weight used in raids and bombing attacks, and the swift fighting machines, quick to rise high, and swift to manoeuvre which would protect the former from the enemy, or drive away the enemy's observation machines as the case might be. In the form which the belligerents finally adopted as most advantageous the fighting aeroplanes were mainly biplanes equipped with powerful motors seldom of less than 140 horse-power, and carrying often but one man who is not merely the pilot, but the operator of the machine gun with which each was equipped. Still planes carrying two men, and even three of whom one was the pilot, the other two the operators of the machine guns were widely adopted. They had indeed their disadvantages. They were slower to rise and clumsier in the turns. The added weight of the two gunmen cut down the amount of fuel that could be carried and limited the radius of action. But one curious disadvantage which would not at first suggest itself to the lay mind

A Caproni triplane showing propellers and fuselage

was the fact that the roar of the propeller was so great that no possible communication could pass between the pilot and the gunner. Their co-operation must be entirely instinctive or there could be no unity of action—and in practice it was found that there was little indeed. The smaller machine, carrying but one man, was quicker in the get-away and could rise higher in less time—a most vital consideration, for in the tactics of aerial warfare it is as desirable to get above your enemy as in the days of the old line of battleships it was advantageous to secure a position off the stern of your enemy so that you might rake him fore and aft.

The machines ultimately found to best meet the needs of aerial fighting were for the Germans always the Fokker, and the Taube—so called from its resemblance to a flying dove, though it was far from being the dove of peace. The wings are shaped like those of a bird and the tail adds to the resemblance. The Allies after testing the Taube design contemptuously rejected it, and indeed the Germans themselves substituted the Fokker for it in the war's later days.

The English used the "Vickers Scout," built of aluminium and steel and until late in the war usually designed to carry two aviators. This machine unlike most of the others has the propeller at the stern, called a "pusher" in contradistinction to the "tractor," acting as the screw of a ship and avoiding the interference with the rifle fire which the pulling, or tractor propeller mounted before the pilot to a certain degree presents. The Vickers machine is lightly armoured. The English also use what was known as the "D. H. 5," a machine carrying a motor of very high horse-power, while the Sopwith and Bristol biplane were popular as fighting craft.

The French pinned their faith mainly to the Farman, the Caudron, the Voisin, and the Moraine-Saulnier machines. The Bleriot and the Nieuport, which were for some reason ruled out at the beginning of the war, were afterwards re-adopted and employed in great numbers.

It would be gratifying to an American author to be able to describe, or at least to mention, the favourite machine of the American aviators who flocked to France immediately upon the declaration of war, but the mortifying fact is that having no aeroplanes of our own, our gallant volunteer soldiers of the air had to be equipped throughout by the French with machines of their favourite types. After we entered the war we adopted a 'plane of American design to which was given the name "Liberty plane."

It may be worth while to revert for a moment to the distinction drawn in a preceding paragraph between the pusher propeller and the tractor which revolved in front of the aviator and of his machine gun. It would seem almost incredible that two heavy blades of hard wood revolving at a speed not less that twelve hundred times a minute, a speed so rapid that their passage in front of the eyes of the aviator interfered in no way with his vision, should not have blocked a stream of bullets falling from a gun at the rate of more than six hundred a minute. Nevertheless it was claimed during the earlier days of the war that these bullets were not appreciably diverted by the whirling propellers nor were the latter apparently injured by the missiles. The latter assertion, however, must have been to some extent disproved because it came about that the propellers of the later machines were rimmed with a thin coating of steel lest the blades be cut by the bullets. But the amazing ability of modern science to cope with what seemed to be an insoluble problem was demonstrated by the invention of a device light and compact enough to be carried in an aeroplane, which applied to the machine gun and timed in accordance with the revolutions of the propeller so synchronized the shots with those revolutions that the stream of lead passed between the whirling blades never once striking. The machine was entirely automatic, requiring no attention on the part of the operator after the gun was once started on its discharge. This device was originally used by the Germans who applied it to their Fokker machines. It was claimed for it that by doing away with the wastage caused by the diversion of the course of bullets, which struck the revolving propellers, it actually saved for effective use about thirty per cent. of the ammunition employed. As the amount of ammunition which can be carried by an aeroplane is rigidly limited this gave to the appliance a positive value.

Reference has been made to the extraordinary immunity of flying aeroplanes to the attacks of anti-aircraft guns. The number of wounds they could sustain without being brought to earth was amazing. Grahame-White tells of a comparison made in one of the aerodromes of the wounds sustained by the machines after a day's hard scouting and fighting. One was found to have been hit no less than thirty-seven times. Curiously enough the man who navigated it escaped unscathed. Wounds in the wings are harmless. But the puncture of the fuel tank almost certainly means an explosion and the death of the aviator in the flame thousands of feet in the air. During an air battle before Arras, a British aviator encountered this fate. When his tank was struck and the fuselage, or body, of his machine burst into flames, he

THE TERROR THAT FLIETH BY NIGHT

knew that he was lost. By no possibility could he reach the ground before he should be burned to death. A neighbouring aviator flying not far from him told the story afterwards:

> Jack was not in the thick of this fight. He was rather on the outskirts striving to get in when I suddenly saw his whole machine enveloped in a sheet of flame. Instantly he turned towards the nearest German and made at him with the obvious intention of running him down and carrying him to earth in the same cloud of fire. The man thus threatened, twisted and turned in a vain effort to escape the red terror bearing down upon him. But suffering acutely as he must have been, Jack followed his every move until the two machines crashed, and whirling over and over each other like two birds in an aerial combat fell to earth and to destruction. They landed inside the German lines so we heard no more about them. But we could see the smoke from the burning debris for some time.

As the range of anti-aircraft guns increased the flyers were driven higher and higher into the air to escape their missiles. At one time 4500 feet was looked upon as a reasonably safe height, but when the war had been under way about two years the weapons designed to combat aircraft were so improved that they could send their shots effectively 10,000 feet into the air. If the aircraft had been forced to operate at that height their usefulness would have been largely destroyed, for it is obvious that for observation purposes the atmospheric haze at such a height would obscure the view and make accurate mapping of the enemy's position impossible. For offensive purposes too the aeroplanes at so great an elevation would be heavily handicapped, if not indeed rendered impotent. As we shall see later, dropping a bomb from a swiftly moving aeroplane upon a target is no easy task. It never falls direct but partakes of the motion of the plane. It is estimated that for every thousand feet of elevation a bomb will advance four hundred feet in the direction that the aircraft is moving, provided its speed is not in excess of sixty miles an hour. As a result marksmanship at a height of more than five thousand feet is practically impossible.

In the main this situation is met, as all situations in war in which efficiency can only be attained at the expense of great personal danger are met, namely, by braving the danger. When the aviators have an attack in contemplation they fly low and snap their fingers at the puff balls of death as the shrapnel from their appearance when bursting may well be called. Naturally, efforts were made early in the war to lessen the danger by armouring the body of the machine sufficiently to protect the aviator and his engine—for if the aviator escaped a shot which found the engine, his plight would be almost as bad as if the missile had struck him.

The main difficulty with armouring the machines grew out of the added weight. The more efficient the armour, the less fuel could be carried and the less ammunition. If too heavily loaded the speed of the machine would be reduced and its ability to climb rapidly upon which the safety of the aviator usually depends, either in reconnaissance or fighting, would be seriously impeded. The first essays in protective armour took the form of the installation of a thin sheet of steel along the bottom of the body of the craft. This turned aside missiles from below provided the plane were not so near the ground as to receive them at the moment of their highest velocity. But it was only an unsatisfactory makeshift. At the higher altitudes it was unnecessary and in conflict with other aeroplanes it proved worthless, because in a battle in the air the shots of the enemy are more likely to come from above or at least from levels in the same plane. The armoured aeroplane was quickly found to have less chance of mounting above its enemy, because of the weight it carried, and before long the principle of protecting an aeroplane as a battleship is protected was abandoned, except in the case of the heavier machines intended to operate as scouts or guides to artillery, holding their flights near the earth and protected from attack from above by their attendant fleet of swift fighting machines. Of these the Vickers machine used mainly by the British is a common type. It is built throughout of steel and aluminium, and the entire fuselage is clothed with steel plating which assures protection to the two occupants from either upward or lateral fire. The sides of the body are carried up so that only the heads of the aviators are visible. But to accomplish this measure of protection for the pilot and the gunner who operates the machine gun from a seat forward of the pilot, the weight of the craft is so greatly increased that it is but little esteemed for any save the most sluggish manoeuvre.

Indeed just as aircraft, as a factor in war, have come to be more like the cavalry in the army, or the destroyers and scout cruisers in the navy, so the tendency has been to discard everything in their design that might by any possibility interfere with their speed and their ability to turn and twist, and change direction and elevation with the utmost celerity under the most difficult of conditions. It is possible that

A Curtis Seaplane Leaving a Battleship

should this war run into the indefinite future we may see aircraft built on ponderous lines and heavily armoured, and performing in the air some of the functions that the British "tanks" have discharged on the battlefields. But at the end of three years of war, and at the moment when aerial hostilities seemed to be engaging more fully than even before the inventive genius of the nations, and the dash and skill of the fighting flyers, the tendency is all toward the light and swift machine.

The attitude of the fighting airmen is somewhat reminiscent of that of America's greatest sea-fighter, Admiral Farragut. Always opposed to ironclads, the hero of Mobile Bay used to say that when he went to sea he did not want to go in an iron coffin, and that when a shell had made its way through one side of his ship he didn't want any obstacle presented to impede its passing out of the other side.

The all important and even vital necessity for speed also detracted much from the value of aircraft in offensive operations. It was found early that you could not mount on a flying machine guns of sufficient calibre to be of material use in attacking fortified positions. If it was necessary for the planes to proceed any material distance before reaching their objective, the weight of the necessary fuel would preclude the carriage of heavy artillery. In the case of seaplanes which might be carried on the deck of a battleship to a point reasonably contiguous to the object to be attacked, this difficulty was not so serious. This was demonstrated to some extent by the British raids on the German naval bases of Cuxhaven and Wilhelmshaven, but even in these instances it was bombs dropped by aviators, not gunfire that injured the enemy's works. But for the aeroplane proper this added weight was so positive a handicap as to practically destroy its usefulness as an assailant of fortified positions.

The heavier weapons of offence which could be carried by the aeroplane even of the highest development were the bombs. These once landed might cause the greatest destruction, but the difficulty of depositing them directly upon a desired target was not to be overcome. The dirigible balloon enjoyed a great advantage over the aeroplane in this respect, for it was able to hover over the spot which it desired to hit and to discharge its bombs in a direct perpendicular line with enough initial velocity from a spring gun to overcome largely any tendency to deviate from the perpendicular. But an aeroplane cannot stop. When it stops it must descend. If it is moving at the moderate speed of sixty miles an hour when it drops its missile, the bomb itself will move forward at the rate of sixty miles an hour until gravity has overcome the

LAUNCHING A HYDRO-AEROPLANE

initial forward force. Years before the war broke out, tests were held in Germany and France of the ability of aviators to drop a missile upon a target marked out upon the ground. One such test in France required the dropping of bombs from a height of 2400 feet upon a target 170 feet long by 40 broad—or about the dimensions of a small and rather stubby ship. The results were uniformly disappointing. The most creditable record was made by an American aviator, Lieutenant Scott, formerly of the United States Army. His first three shots missed altogether, but thereafter he landed eight within the limits. In Germany the same year the test was to drop bombs upon two targets, one resembling a captive Zeppelin, the other a military camp 330 feet square. The altitude limit was set at 660 feet. This, though a comparatively easy test, was virtually a failure. Only two competitors succeeded in dropping a bomb into the square at all, while the balloon was hit but once.

The character and size of the bombs employed by aircraft naturally differed very widely, particularly as to size, between those carried by dirigibles and those used by aeroplanes. The Zeppelin shell varied in weight between two hundred and two hundred and fifty pounds. It was about forty-seven inches long by eight and a half inches in diameter. Its charge varied according to the use to which it was to be put. If it was hoped that it would drop in a crowded spot and inflict the greatest amount of damage to human life and limb it would carry a bursting charge, shrapnel, and bits of iron, all of which on the impact of the missile upon the earth would be hurled in every direction to a radius exceeding forty yards. If damage to buildings, on the other hand, was desired, some high explosive such as picric acid would be used which would totally wreck any moderate-sized building upon which the shell might fall. In many instances, particularly in raids upon cities such as London, incendiary shells were used charged with some form of liquid fire, which rapidly spread the conflagration, and which itself was practically inextinguishable.

Shells or bombs of these varying types were dropped from aeroplanes as well as from the larger and steadier Zeppelins. The difference was entirely in the size. It was said that a Zeppelin might drop a bomb of a ton's weight. But so far as attainable records are concerned it is impossible to cite any instance of this being done. The effect on the great gas bag of the sudden release of a load so great would certainly cause a sudden upward flight which might be so quick and so powerful as to affect the very structure of the ship. So far as known 250 pounds was the topmost limit of Zeppelin bombs, while most of them were of much smaller dimensions. The aeroplane bombs were seldom more than sixty pounds in weight, although in the larger British machines a record of ninety-five pounds has been attained. The most common form of bomb used in the heavier-than-air machines was pear-shaped, with a whirling tail to keep the missile upright as it falls. Steel balls within, a little larger than ordinary shrapnel, are held in place by a device which releases them during the fall. On striking the ground they fall on the explosive charge within and the shell bursts, scattering the two or three hundred steel bullets which it carries over a wide radius. Bombs of this character weigh in the neighbourhood of six pounds and an ordinary aeroplane can carry a very considerable number. Their exploding device is very delicate so that it will operate upon impact with water, very soft earth, or even the covering of an airship. Other bombs commonly used in aeroplanes were shaped like darts, winged like an arrow so that they would fall perpendicularly and explode by a pusher at the point which was driven into the body of the bomb upon its impact with any hard substance.

It seems curious to read of the devices sometimes quite complicated and at all times the result of the greatest care and thought, used for dropping these bombs. In the trenches men pitched explosive missiles about with little more care than if they had been so many baseballs, but only seldom was a bomb from aloft actually delivered by hand. In the case of the heavier bombs used by the dirigibles this is understandable. They could not be handled by a single man without the aid of mechanical devices. Some are dropped from a cradle which is tilted into a vertical position after the shell has been inserted. Others are fired from a tube not unlike the torpedo tube of a submarine, but which imparts only slight initial velocity to the missile. Its chief force is derived from gravity, and to be assured of its explosion the aviator must discharge it from a height proportionate to its size.

In the aeroplane the aviator's methods are more simple. Sometimes the bombs are carried in a rack beneath the body of the machine, and released by means of a lever at the side. A more primitive method often in use is merely to attach the bomb to a string and lower it to a point at which the aviator is certain that in falling it will not touch any part of the craft, and then cut the string. Half a dozen devices by which the aviator can hold the bomb at arm's length and drop it with the certainty of a perpendicular fall are in use in the different air navies. It will be evident to the most casual consideration that with any one of these devices employed by an aviator in a machine going at a speed of sixty miles an hour or more the matter of hitting the target is one in which luck has a very great share.

There is good reason for the pains taken by the aviators to see that their bombs fall swift and true, and clear of all the outlying parts of their machines. The grenadier in the trenches has a clear field for his explosive missile and he may toss it about with what appears to be desperate carelessness—though instances have been known in which a bomb thrower, throwing back his arm preparatory to launching his canned volcano, has struck the back of his own trench with disastrous results. But the aviator must be even more careful. His bombs must not hit any of the wires below his machine in falling—else there will be a dire fall for him. And above all they must not get entangled in stays or braces. In such case landing will bring a most unpleasant surprise.

A striking case was that of a bomber who had been out over the German trenches. He had a two-man machine, had made a successful flight and had dropped, effectively as he supposed, all his bombs. Returning in serene consciousness of a day's duty well done, he was about to spiral down to the landing place when his passenger looked over the side of the car to see if everything was in good order. Emphatically it was not. To his horror he discovered that two of the bombs had not fallen, but had caught in the running gear of his machine. To attempt a landing with the bombs in this position would have been suicidal. The bombs would have instantly exploded, and annihilated both machine and aviators. But to get out of the car, climb down on the wires, and try to unhook the bombs seemed more desperate still. Stabilizers, and other devices, now in common use, had not then been invented and to go out on the wing of a biplane, or to disturb its delicate balance, was unheard of. Nevertheless it was a moment for desperate remedies. The pilot clung to his controls, and sought to meet the shifting strains, while the passenger climbed out on the wing and then upon the running gear. To trust yourself two thousand feet in mid-air with your feet on one piano wire, and one hand clutching another, while with the other hand you grope blindly for a bomb charged with high explosive, is an experience for which few men would yearn. But in this case it was successful. The bombs fell—nobody cared where—and the two imperilled aviators came to ground safely.

A form of offensive weapon which for some reason seems peculiarly horrible to the human mind is the *fléchette*. These are steel darts a little larger than a heavy lead pencil and with the upper two thirds of the stem deeply grooved so that the greater weight of the lower

AT A UNITED STATES TRAINING CAMP

part will cause them to fall perpendicularly. These are used in attacks upon dense bodies of troops. Particularly have they proved effective in assailing cavalry, for the nature of the wounds they produce invariably maddens the horses who suffer from them and causes confusion that will often bring grave disaster to a transport or artillery train. Though very light, these arrows when dropped from any considerable height inflict most extraordinary wounds. They have been known to penetrate a soldier's steel helmet, to pass through his body and that of the horse he bestrode, and bury themselves in the earth. In the aeroplane they are carried in boxes of one hundred each, placed over an orifice in the floor. A touch of the aviator's foot and all are discharged. The speed of the machine causes them to fall at first in a somewhat confused fashion, with the result that before all have finally assumed their perpendicular position they have been scattered over a very considerable extent of air. Once fairly pointed downward they fall with unerring directness points downward to their mark.

It is a curious fact that not long after these arrows first made their appearance in the French machines, they were imitated by the Germans, but the German darts had stamped upon them the words: "Made in Germany, but invented by the French."

One of the duties of the fighting airmen is to destroy the observation balloons which float in great numbers over both the lines tugging lazily at the ropes by which they are held captive while the observers perched in their baskets communicate the results of their observations by telephone to staff officers at a considerable distance. These balloons are usually anchored far enough back of their own lines to be safe from the ordinary artillery fire of their enemies. They were therefore fair game for the mosquitoes of the air. But they were not readily destroyed by such artillery as could be mounted on an ordinary aeroplane. Bullets from the machine-guns were too small to make any rents in the envelope that would affect its stability. Even if incendiary they could not carry a sufficiently heavy charge to affect so large a body. The skin of the "sausages," as the balloons were commonly called from their shape, was too soft to offer sufficient resistance to explode a shell of any size. The war was pretty well under way before the precise weapon needed for their destruction was discovered. This proved to be a large rocket of which eight were carried on an aeroplane, four on each side. They were discharged by powerful springs and a mechanism started which ignited them as soon as they had left the aeroplane behind. The head of each rocket was of pointed steel, very sharp and heavy

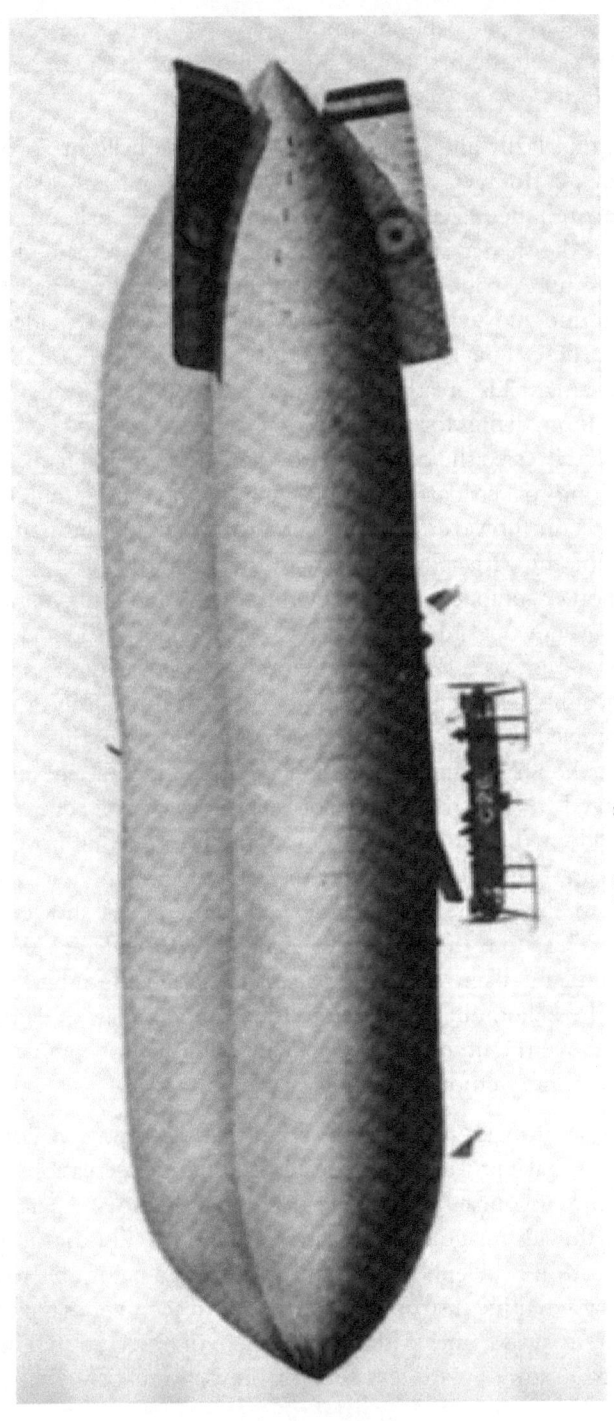

A "BLIMP" WITH GUN MOUNTED ON TOP

enough to pierce the balloon skin. Winslow was fortunate enough to be present when the first test of this weapon was made. In his book, *With the French Flying Corps*, he thus tells the story:

> Swinging lazily above the field was a captive balloon. At one end of Le Bourget was a line of waiting aeroplanes. "This is the second; they have already brought down one balloon," remarked the man at my elbow. The hum of a motor caused me to look up. A wide-winged double motor, Caudron, had left the ground and was mounting gracefully above us. Up and up it went, describing a great circle, until it faced the balloon. Everyone caught his breath. The Caudron was rushing straight at the balloon, diving for the attack.
>
> "Now!" cried the crowd. There was a loud crack, a flash, and eight long rockets darted forth leaving behind a fiery trail. The aviator's aim however was wide, and to the disappointment of everyone the darts fell harmlessly to the ground.
>
> Another motor roared far down the field, and a tiny *appareil de chasse* shot upward like a swallow. "A Nieuport," shouted the crowd as one voice. Eager to atone for his *copain's* failure, and impatient at his delay in getting out of the way, the tiny biplane tossed and tumbled about in the air like a clown in the circus ring.
>
> "Look! he's looping! he falls! he slips! no, he rights again!" cried a hundred voices as the skilful pilot kept our nerves on edge.
>
> Suddenly he darted into position and for a second hovered uncertain. Then with a dive like that of a dragon-fly, he rushed down to the attack. Again a sheet of flame and a shower of sparks. This time the balloon sagged. The flames crept slowly around its silken envelope. *"Touchez!"* cried the multitude. Then the balloon burst and fell to the ground a mass of flames. High above the little Nieuport saucily continued its pranks, as though contemptuous of such easy prey.

It may be properly noted at this point that the captive balloons or kite balloons have proved of the greatest value for observations in this war. Lacking of course the mobility of the swiftly moving aeroplanes, they have the advantage over the latter of being at all times in direct communication by telephone with the ground and being able to carry quite heavy scientific instruments for the more accurate mapping out of such territory as comes within their sphere of observation. They are not easy to destroy by artillery fire, for the continual swaying of the balloon before the wind perplexes gunners in their aim. At a height of six hundred feet, a normal observation post, the horizon is nearly thirty miles from the observer. In flat countries like Flanders, or at sea where the balloon may be sent up from the deck of a ship, this gives an outlook of the greatest advantage to the army or fleet relying upon the balloon for its observations of the enemy's dispositions.

Most of the British and French observation balloons have been of the old-fashioned spherical form which officers in those services find sufficiently effective. The Germans, however, claimed that a balloon might be devised which would not be so very unstable in gusty weather. Out of this belief grew the Parseval-Siegfeld balloon which from its form took the name of the Sausage. In fact its appearance far from being terrifying suggests not only that particular edible, but a large dill pickle floating awkwardly in the air. In order to keep the balloon always pointed into the teeth of the wind there is attached to one end of it a large surrounding bag hanging from the lower half of the main envelope. One end of this, the end facing forward, is left open and into this the wind blows, steadying the whole structure after the fashion of the tail of a kite. The effect is somewhat grotesque as anyone who has studied the numerous pictures of balloons of this type employed during the war must have observed. It looks not unlike some form of tumour growing from a healthy structure.

Captive or kite balloons are especially effective as coast guards. Posted fifty miles apart along a threatened coast they can keep a steady watch over the sea for more than twenty-five miles toward the horizon. With their telephonic connections they can notify aeroplanes in waiting, or for that matter swift destroyers, of any suspicious sight in the distance, and secure an immediate investigation which will perhaps result in the defeat of some attempted raid. Requiring little power for raising and lowering them and few men for their operation, they form a method of standing sentry guard at a nation's front door which can probably be equalled by no other device. The United States at the moment of the preparation of this book is virtually without any balloons of this type—the first one of any pretensions having been tested in the summer of 1917.

As late as the third year of the war it could not be said that the possibilities of aerial offense had been thoroughly developed by any nation. The Germans indeed had done more than any of the belligerents in this direction with their raids on the British coast and on London. But, as already pointed out, these raids as serious attacks on strategic

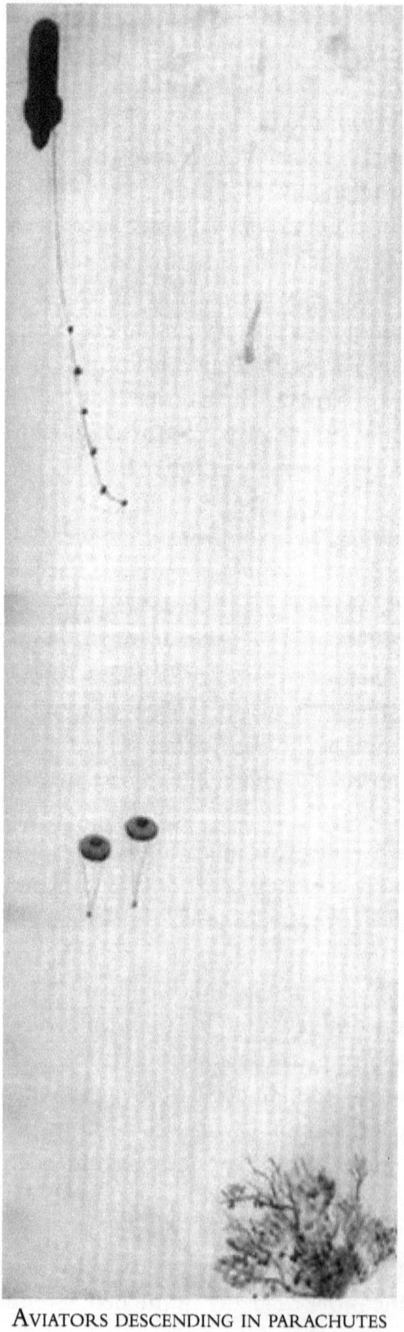

AVIATORS DESCENDING IN PARACHUTES
FROM A BALLOON STRUCK BY
INCENDIARY SHELLS

THE BALLOON FROM WHICH THE
AVIATORS FLED

positions were mere failures. Advocates of the increased employment of aircraft in this fashion insist that the military value to Germany of the raids lay not so much in the possibility of doing damage of military importance but rather in the fact that the possibility of repeated and more effective raids compelled Great Britain to keep at home a force of thirty thousand to fifty thousand men constantly on guard, who but for this menace would have been employed on the battlefields of France. In this argument there is a measure of plausibility. Indeed between January, 1915, and June 13, 1917, the Germans made twenty-three disastrous raids upon England, killing more than seven hundred persons and injuring nearly twice as many. The amount of damage to property has never been reported nor is it possible to estimate the extent of injury inflicted upon works of a military character. The extreme secrecy with which Great Britain, in common with the other belligerents, has enveloped operations of this character makes it impossible at this early day to estimate the military value of these exploits. Merely to inflict anguish and death upon a great number of civilians, and those largely women and children, is obviously of no military service. But if such suffering is inflicted in the course of an attack which promises the destruction or even the crippling of works of military character like arsenals, munitions plants, or naval stores, it must be accepted as an incident of legitimate warfare. The limited information obtainable in wartime seems to indicate that the German raids had no legitimate objective in view but were undertaken for the mere purpose of frightfulness.

The methods of defence employed in Great Britain, where all attacks must come from the sea, were mainly naval. What might be called the outer, or flying, defences consisted of fast armed fighting seaplanes and dirigibles. Stationed on the coast and ready on the receipt of a wireless warning from scouts, either aerial or naval, that an enemy air flotilla was approaching the coast, they could at once fly forth and give it battle. A thorough defence of the British territory demanded that the enemy should be driven back before reaching the land. Once over British territory the projectiles discharged whether by friend or foe did equal harm to the people on the ground below. Accordingly every endeavour was made to meet and beat the raiders before they had passed the barrier of sea. Beside the flying defences there were the floating defences. Anti-aircraft guns were mounted on different types of ships stationed far out from the shore and ever on the watch. But these latter were of comparatively little avail, for flying over the Channel or the North Sea the invaders naturally flew at a great height. They had no targets there to seek, steered by their compasses, and were entirely indifferent to the prospect beneath them. Moreover anti-aircraft guns, hard to train effectively from an immovable mount, were particularly untrustworthy when fired from the deck of a rolling and tossing ship in the turbulent Channel.

Third in the list of defences of the British coast, or of any other coast which may at any time be threatened with an aerial raid, are defensive stations equipped not only with anti-aircraft guns and search-lights but with batteries of strange new scientific instruments like the "listening towers," equipped with huge microphones to magnify the sound of the motors of approaching aircraft so that they would be heard long before they could be seen, range finders, and other devices for the purpose of gauging the distance and fixing the direction of an approaching enemy.

Some brief attention may here be given to the various types of anti-aircraft guns. These differ very materially in type and weight in the different belligerent armies and navies. They have but one quality in common, namely that they are most disappointing in the results attained. Mr. F. W. Lancaster, the foremost British authority on aircraft, says on this subject: "Anti-aircraft firing is very inaccurate, hence numbers of guns are employed to compensate."

That is to say that one or two guns can be little relied upon to put a flyer *hors du combat*. The method adopted is to have large batteries which fairly fill that portion of the air through which the adventurous airman is making his way with shells fired rather at the section than at the swiftly moving target.

"Archibald," the British airmen call, for some mysterious reason, the anti-aircraft guns employed by their enemies, sometimes referring to a big howitzer which made its appearance late in the war as "Cuthbert." The names sound a little effeminate, redolent somehow of high teas and the dancing floor, rather than the field of battle. Perhaps this was why the British soldiers adopted them as an expression of contempt for the enemy's batteries. But contempt was hardly justifiable in face of the difficulty of the problem. A gun firing a twenty-pound shrapnel shell is not pointed on an object with the celerity with which a practised revolver shot can throw his weapon into position. The gunner on the ground seeing an aeroplane flying five thousand feet above him—almost a mile up in the air—hurries to get his piece into position for a shot. But while he is aiming the

German air raiders over England: In the foreground three British planes are advancing to the attack

flyer, if a high-speed machine, will be changing its position at a rate of perhaps 120 miles an hour. Nor does it fly straight ahead. The gunner cannot point his weapon some distance in advance as he would were he a sportsman intent on cutting off a flight of wild geese. The aviator makes quick turns—zigzags—employs every artifice to defeat the aim of his enemy below. Small wonder that in the majority of cases they have been successful. The attitude of the airmen toward the "Archies" is one of calm contempt.

The German mind being distinctly scientific invented early in the war a method of fixing the range and position of an enemy aeroplane which would be most effective if the target were not continually in erratic motion. The method was to arrange anti-aircraft guns in a triangle, all in telephonic connection with a central observer. When a flyer enters the territory which these guns are guarding, the gunner at one of the apexes of the triangle fires a shell which gives out a red cloud of smoke. Perhaps it falls short. The central observer notes the result and orders a second gun to fire. Instantly a gunner at another apex fires again, this time a shell giving forth black smoke. This shell discharged with the warning given by the earlier one is likely to come nearer the target, but at any rate marks another point at which it has been missed. Between the two a third gunner instantly corrects his aim by the results of the first two shots. His shell gives out a yellow smoke. The observer then figures from the positions of the three guns the lines of a triangular cone at the apex of which the target should be. Sometimes science wins, often enough for the Germans to cling to the system. But more often the shrewd aviator defeats science by his swift and eccentric changes of his line of flight.

At the beginning of the war Germany was very much better equipped with anti-aircraft guns than any of her enemies. This was due to the remarkable foresight of the great munitions makers, Krupp and Ehrhardt, who began experimenting with anti-aircraft guns before the aircraft themselves were much more than experiments. The problem was no easy one. The gun had to be light, mobile, and often mounted on an automobile so as to be swiftly transferred from place to place in pursuit of raiders. It was vital that it should be so mounted as to be speedily trained to any position vertical or horizontal. As a result the type determined upon was mounted on a pedestal fixed to the chassis of an automobile or to the deck of a ship in case it was to be used in naval warfare. The heaviest gun manufactured in Germany was of 4-1/4-inch calibre, throwing a shell of forty pounds weight. This could be mounted directly over the rear axle of a heavy motor truck. To protect the structure of the car from the shock of the recoil these guns are of course equipped with hydraulic or other appliances for taking it up. They are manufactured also in the 3-inch size. Germany, France, and England vied with each other in devising armoured motor cars equipped with guns of this type—the British using the makes of Vickers and Hotchkiss, and the French their favourite Creusot. The trucks are always armoured, the guns mounted in turrets so that the effect is not unlike that of a small battleship dashing madly down a country road and firing repeatedly at some object directly overhead. But the record has not shown that the success of these picturesque and ponderous engines of war has been great. They cannot manoeuvre with enough swiftness to keep up with the gyrations of an aeroplane. They offer as good a target for a bomb from above as the aircraft does to their shots from below. Indeed they so thoroughly demonstrated their inefficiency that before the war had passed its third year they were either abandoned or their guns employed only when the car was stationary. Shots fired at full speed were seldom effective.

The real measure of the effectiveness of anti-aircraft guns may be judged by the comparative immunity that attended the aviators engaged on the two early British raids on Friedrichshaven, the seat of the great Zeppelin works on Lake Constance, and on the German naval base at Cuxhaven. The first was undertaken by three machines. From Belfort in France, the aviators turned into Germany and flew for 120 miles across hostile territory. The flight was made by day though indeed the adventurous aviators were favoured by a slight mist. Small single seated "avro" machines were used, loaded heavily with bombs as well as with the large amount of fuel necessary for a flight which before its completion would extend over 250 miles. Not only at the frontier, but at many fortified positions over which they passed, they must have exposed themselves to the fire of artillery, but until they actually reached the neighbourhood of the Zeppelin works they encountered no fire whatsoever. There the attack on them was savage and well maintained. On the roofs of the gigantic factory, on neighbouring hillocks and points of vantage there were anti-aircraft guns busily discharging shrapnel at the invaders. It is claimed by the British that fearing this attack the Germans had called from the front in Flanders their best marksmen, for at that time the comparative worthlessness of the Zeppelin had not been demonstrated and the protection of the works was regarded as a prime duty of the army.

One aviator's narrow escape

The invading machines flew low above the factory roofs. The adventurers had come far on an errand which they knew would awaken the utmost enthusiasm among their fellows at home and they were determined to so perform their task that no charge of having left anything undone could possibly lie. Commander Briggs, the first of the aviators to reach the scene, flew as low as one hundred feet above the roofs, dropping his bombs with deadly accuracy. But he paid for his temerity with the loss of his machine and his liberty. A bullet pierced his petrol tank and there was nothing for him to do save to glide to earth and surrender. The two aviators who accompanied him although their machines were repeatedly hit were nevertheless able to drop all their bombs and to fly safely back to Belfort whence they had taken their departure some hours before. The measure of actual damage done in the raid has never been precisely known. Germany always denied that it was serious, while the British ascribe to it the greatest importance—a clash of opinion common in the war and which will for some years greatly perplex the student of its history.

The second raid, that upon Cuxhaven, was made by seaplanes so far as the air fighting was concerned, but in it not only destroyers but submarines also took part. It presented the unique phenomenon of a battle fought at once above, upon, and below the surface of the sea. It is with the aerial feature of the battle alone that we have to do.

Christmas morning, 1915, seven seaplanes were quietly lowered to the surface of the water of the North Sea from their mother ships a little before daybreak. The spot was within a few miles of Cuxhaven and the mouth of the River Elbe. As the aircraft rose from the surface of the water and out of the light mist that lay upon it, they could see in the harbour which they threatened, a small group of German warships. Almost at the same moment their presence was detected. The alarms of the bugles rang out from the hitherto quiet craft and in a moment with the smoke pouring from their funnels destroyers and torpedo boats moved out to meet the attack. Two Zeppelins rose high in the air surrounded by a number of the smaller aeroplanes, eager for the conflict. The latter proceeded at once to the attack upon the raiding air fleet, while the destroyers, the heavier Zeppelins, and a number of submarines sped out to sea to attack the British ships. The mist, which grew thicker, turned the combat from a battle into a mere disorderly raid, but out of it the seaplanes emerged unhurt. All made their way safely back to the fleet, after having dropped their bombs with a degree of damage never precisely known. The weakness of the seaplane is that on returning to its parent ship it cannot usually alight upon her deck, even though a landing platform has been provided. It must, as a rule, drop to the surface of the ocean, and if this be at all rough the machine very speedily goes to pieces. This was the case with four of the seven seaplanes which took part in the raid on Cuxhaven. All however delivered their pilots safely to the awaiting fleet and none fell a victim to the German anti-aircraft guns.

In May of 1917, the British Royal Naval Air Service undertook the mapping of the coast of Belgium north from Nieuport, the most northerly seaport held by the British, to the southern boundary of Holland. This section of coast was held by the Germans and in it were included the two submarine bases of Zeebrugge and Ostend. At the latter point the long line of German trenches extending to the boundary of Switzerland rested its right flank on the sea. The whole coast north of that was lined with German batteries, snugly concealed in the rolling sand dunes and masked by the waving grasses of a barren coast. From British ships thirty miles out at sea, for the waters there are shallow and large vessels can only at great peril approach the shore, the seaplanes were launched. Just south of Nieuport a land base was established as a rendezvous for both air-and seaplanes when their day's work was done. From fleet and station the aerial observers took their way daily to the enemy's coast. Every mile of it was photographed. The hidden batteries were detected and the inexorable record of their presence imprinted on the films. The work in progress at Ostend and Zeebrugge, the active construction of basins, locks, and quays, the progress of the great mole building at the latter port, the activities of submarines and destroyers within the harbour, the locations of guns and the positions of barracks were all indelibly set down. These films developed at leisure were made into coherent wholes, placed in projecting machines, and displayed like moving pictures in the ward rooms of the ships hovering off shore, so that the naval forces preparing for the assault had a very accurate idea of the nature of the defences they were about to encounter.

This was not done of course without considerable savage fighting in mid-air. The Germans had no idea of allowing their defences and the works of their submarine bases to be pictured for the guidance of their foes. Their anti-aircraft guns barked from dawn to dark whenever a British plane was seen within range. Their own aerial fighters were continually busy, and along that desolate wave-washed coast many a lost lad in leather clothing and goggles, crumpled up in the ruins of his machine after a fall of thousands of feet, lay as a memorial to the prowess of the defenders of the coast and the audacity of those who sought to invade it. But during the long weeks of this extended reconnaissance hardly a spadeful of dirt could be moved, a square yard of concrete placed in position, or a submarine or torpedo boat manoeuvred without its record being entered upon the detailed charts the British were so painstakingly preparing against the day of assault. When peace shall finally permit the publication of the records of the war, now held secret for military reasons, such maps as those prepared by the British air service on the Belgian coast will prove most convincing evidence of the military value of the aerial scouts.

What the lads engaged in making these records had to brave in the way of physical danger is strikingly shown by the description of a combat included in one of the coldly matter-of-fact official reports. The battle was fought at about twelve thousand feet above mother earth. We quote the official description accompanied by some explanatory comments added by one who was an eye-witness and who conversed with the triumphant young airman on his return to the safety of the soil.

"While exposing six plates," says the official report of this youthful recording angel, "I observed five H. A.'s cruising."

"H. A." stands for "hostile aeroplane."

"Not having seen the escort since returning inland, the pilot prepared to return. The enemy separated, one taking up a position above the tail and one ahead. The other three glided toward us on the port side, firing as they came. The two diving machines fired over 100 rounds, hitting the pilot in the shoulder."

As a matter of fact, the bullet entered his shoulder from above, behind, breaking his left collarbone, and emerged just above his heart, tearing a jagged rent down his breast. Both his feet, furthermore, were pierced by bullets; but the observer is not concerned with petty detail.

The observer held his fire until H. A., diving on tail, was within five yards. Here it might be mentioned that the machines were hurtling through space at a speed in the region of one hundred miles an hour.

The pilot of H. A., having swooped to within speaking distance, pushed up his goggles, and laughed triumphantly as he took sight for the shot that was to end the fight. But the observer, had his own idea how the fight should end.

"I then shot one tray into the enemy pilot's face," he says, with curt relish, "and watched him sideslip and go spinning earthward in a train of smoke."

He then turned his attention to his own pilot. The British machine was barely under control, but as the observer rose in his seat to investigate the foremost gun was fired, and the aggressor ahead went out of control and dived nose first in helpless spirals.

Suspecting that his mate was badly wounded in spite of this achievement, the observer swung one leg over the side of the fuselage and climbed on to the wing—figure for a minute the air pressure on his body during this gymnastic feat—until he was beside the pilot, faint and drenched with blood, who had nevertheless got his machine back into complete control.

"Get back, you ass!" he said through white lips in response to inquiries how he felt. So the ass got back the way he came, and looked around for the remainder of the H. A.'s. These, however, appeared to have lost stomach for further fighting and fled.

The riddled machine returned home at one hundred knots while the observer, having nothing better to do, continued to take photographs.

"The pilot, though wounded, made a perfect landing"—thus the report concludes.

When the time came for the assault upon Zeebrugge the value of these painstaking preparations was made evident. The attack was made from sea and air alike. Out in the North Sea the great British battleships steamed in as near the coast as the shallowness of the water would permit. From the forward deck of each rose grandly a seaplane until the air was darkened by their wings, and they looked like a monstrous flock of the gulls which passengers on ocean-going liners watch wheeling and soaring around the ship as it ploughs its way through the ocean. These gulls though were birds of prey. They were planes of the larger type, biplanes or triplanes carrying two men, usually equipped with two motors and heavily laden with high explosive bombs. As they made their way toward the land they were accompanied by a fleet of light draft monitors especially built for this service, each mounting two heavy guns and able to manoeuvre in shallow water. With them advanced a swarm of swift, low-lying, dark-painted destroyers ready to watch out for enemy torpedo boats or submarines. They mounted anti-aircraft guns too and were prepared to defend the monitors against assaults from the heavens above as well as from the sinister attack of the underwater boats. Up from the land base at

Nieuport came a great fleet of aeroplanes to co-operate with their naval brethren. Soon upon the German works, sheltering squadrons of the sinister undersea boats, there rained a hell of exploding projectiles from sea and sky. Every gunner had absolute knowledge of the precise position and range of the target to which he was assigned. The great guns of the monitors roared steadily and their twelve and fourteen-inch projectiles rent in pieces the bomb proofs of the Germans, driving the Boches to cover and reducing their works to mere heaps of battered concrete. Back and forth above flew seaplanes and aeroplanes, giving battle to the aircraft which the Germans sent up in the forlorn hope of heading off that attack and dropping their bombs on points carefully mapped long in advance. It is true that the aim of the aviators was necessarily inaccurate. That is the chief weakness of a bombardment from the sky. But what was lacking in individual accuracy was made up by the numbers of the bombing craft. One might miss a lock or a shelter, but twenty concentrating their fire on the same target could not all fail. This has become the accepted principle of aerial offensive warfare. The inaccuracy of the individual must be corrected by the multiplication of the number of the assailants.

The attack on Zeebrugge was wholly successful. Though the Germans assiduously strove to conceal the damage done, the later observations of the ruined port by British airmen left no doubt that as a submarine base it had been put out of commission for months to come. The success of the attack led to serious discussion, in which a determination has not yet been reached, of the feasibility of a similar assault upon Heligoland, Kiel, or Cuxhaven, the three great naval bases in which the German fleet has lurked in avoidance of battle with the British fleet. Many able naval strategists declared that it was time for the British to abandon the policy of a mere blockade and carry out the somewhat rash promise made by Winston Churchill when First Lord of the Admiralty, to "dig the rats out of their holes." Such an attack it was urged should be made mainly from the air, as the land batteries and sunken mines made the waters adjacent to these harbours almost impassable to attacking ships. Rear-Admiral Fiske, of the United States Navy, strongly urging such an attack, wrote in an open letter:

> The German Naval General Staff realizes the value of concentration of power and mobility in as large units as possible. The torpedo plane embodies a greater concentration of power

and mobility than does any other mechanism. For its cost, the torpedo plane is the most powerful and mobile weapon which exists at the present day.

An attack by allied torpedo planes, armed with guns to defend themselves from fighting aeroplanes, would be a powerful menace to the German fleet and, if made in sufficient numbers, would give the Allies such unrestricted command of the North Sea, even of the shallow parts near the German coast, that German submarines would be prevented from coming from a German port, the submarine menace abolished, and all chance of German success wiped out.

I beg also to point out that an inspection of the map of Europe shows that in the air raids over land the strategical advantage lies with Germany, because her most important towns, like Berlin, are farther inland than the most important towns of the Allies, like London, so that aeroplanes of the Allies, in order to reach Berlin, would have to fly over greater distances, while exposed to the fire of other aeroplanes, than do aeroplanes of the Germans in going to London for raids on naval vessels.

However, the strategical advantage over water lies with the British, because their control of the deep parts of the North Sea enables them to establish a temporary aeronautical base of mother ships sufficiently close to the German fleet to enable the British to launch a torpedo-plane attack from it on the German fleets in Kiel and Wilhelmshaven, while the Germans could not possibly establish an aeronautical base sufficiently close to the British fleet.

This gives the Allies the greatest advantage of the offensive. It would seem possible, provided a distinct effort is made, for the Allies to send a large number of aeroplane mother ships to a point, say, fifty miles west of Heligoland, and for a large force of fighting aeroplanes and torpedo planes to start from this place about two hours before dawn, reach Kiel Bay and Wilhelmshaven about dawn, attack the German fleets there and sink the German ships.

The distance from Heligoland to Kiel is about ninety land miles, and to Wilhelmshaven about forty-five.

The torpedo planes referred to are an invention of Admiral Fiske's which, in accordance with what seems to be a fixed and fatal prec-

Downed in the enemy's country

edent in the United States, has been ignored by our own authorities but eagerly adopted by the naval services of practically all the belligerents. One weakness of the aerial attack upon ships of war is that the bombs dropped from the air, even if they strike the target, strike upon the protective deck which in most warships above the gunboat class is strong enough to resist, or at least to minimize, the effect of any bomb capable of being carried by an aeroplane. The real vulnerable part of a ship of war is the thin skin of its hull below water and below the armour belt. This is the point at which the torpedo strikes. Admiral Fiske's device permits an aeroplane to carry two torpedoes of the regular Whitehead class and to launch them with such an impetus and at such an angle that they will take the water and continue their course thereunder exactly as though launched from a naval torpedo tube. His idea was adopted both by Great Britain and Germany. British torpedo planes thus equipped sank four Turkish ships in the Sea of Marmora, a field of action which no British ship could have reached after the disastrous failure to force the Dardanelles. The Germans by employment of the same device sank at least two Russian ships in the Baltic and one British vessel in the North Sea. The blindness of the United States naval authorities to the merits of this invention was a matter arousing at once curiosity and indignation among observers during the early days of our entrance upon the war.

Chapter 8
Incidents of the War in the Air

In time, no doubt, volumes will be written on the work of the airmen in the Great War. Except the submarine, no such novel and effective device was introduced into the conduct of this colossal struggle as the scouting aeroplane. The development of the service was steady from the first day when the Belgian flyers proved their worth at Liège. From mere observation trips there sprang up the air duels, from the duels developed skirmishes, and from these in time pitched battles in which several hundred machines would be engaged on each side. To this extent of development aerial tactics had proceeded by midsummer of 1917. Their further development must be left to some future chronicler to record. It must be noted, however, that at that early day the Secretary of the Treasury of the United States, pleading for a larger measure of preparation for the perils of war, asserted that the time was not far distant when this country would have to prepare to repel invading fleets of aircraft from European shores. This may have been an exaggeration. At that moment no aircraft had crossed the Atlantic and no effort to make the passage had been made save those of Wellman and Vanniman. When the guns began to roar on the Belgian frontier there was floating on Keuka Lake, New York, a huge hydro-aeroplane with which it was planned to make the trans-Atlantic voyage. The project had been financed by Mr. Rodman Wanamaker, of Philadelphia, and the tests of the ship under the supervision of a young British army officer who was to make the voyage were progressing most promisingly. But the event that plunged the world into war put a sudden end to experiments like this for the commercial development of the aeroplane. There is every reason to believe, however, that such a flight is practicable and that it will ultimately be made not long after the world shall have returned to peace and sanity.

Later type of French scout: the gun mounted on the upper wing is aimed by pointing the machine and is fired by the pilot

Airmen are not, as a rule, of a romantic or a literary temperament. Pursuing what seems to the onlooker to be the most adventurous and exhilarating of all forms of military service, they have been chary of telling their experiences and singularly set upon treating them as all in the day's work and eliminating all that is picturesque from their narratives. Sergeant James R. McConnell, one of the Americans in the French flying corps, afterwards killed, tells of a day's service in his most readable book, *Flying for France*, in a way that gives some idea of the daily routine of an operator of an *avion de chasse*. He is starting just as the sky at dawn is showing a faint pink toward the eastern horizon, for the aviator's work is best done in early morning when, as a rule, the sky is clear and the wind light:

> Drawing forward out of line, you put on full power, race across the grass, and take the air. The ground drops as the hood slants up before you and you seem to be going more and more slowly as you rise. At a great height you hardly realize you are moving. You glance at the clock to note the time of your departure, and at the oil gauge to see its throb. The altimeter registers 650 feet. You turn and look back at the field below and see others leaving.
>
> In three minutes you are at about four thousand feet. You have been making wide circles over the field and watching the other machines. At forty-five hundred feet you throttle down and wait on that level for your companions to catch up. Soon the escadrille is bunched and off for the lines. You begin climbing again, gulping to clear your ears in the changing pressure. Surveying the other machines, you recognize the pilot of each by the marks on its side—or by the way he flies.
>
> The country below has changed into a flat surface of vari-coloured figures. Woods are irregular blocks of dark green, like daubs of ink spilled on a table; fields are geometrical designs of different shades of green and brown, forming in composite an ultra-cubist painting; roads are thin white lines, each with its distinctive windings and crossings—from which you determine your location. The higher you are the easier it is to read.
>
> In about ten minutes you see the Meuse sparkling in the morning light, and on either side the long line of sausage-shaped observation balloons far below you. Red-roofed Verdun springs into view just beyond. There are spots in it where no red shows and you know what has happened there. In the green

Position of gunner in early French machines

pasture land bordering the town, round flecks of brown indicate the shell holes. You cross the Meuse.

Immediately east and north of Verdun there lies a broad, brown band. From the Woevre plain it runs westward to the "S" bend in the Meuse, and on the left bank of that famous stream continues on into the Argonne Forest. Peaceful fields and farms and villages adorned that landscape a few months ago—when there was no Battle of Verdun. Now there is only that sinister brown belt, a strip of murdered Nature. It seems to belong to another world. Every sign of humanity has been swept away. The woods and roads have vanished like chalk wiped from a blackboard; of the villages nothing remains but grey smears where stone walls have tumbled together. The great forts of Douaumont and Vaux are outlined faintly, like the tracings of a finger in wet sand. One cannot distinguish any one shell crater, as one can on the pockmarked fields on either side. On the brown band the indentations are so closely interlocked that they blend into a confused mass of troubled earth. Of the trenches only broken, half-obliterated links are visible.

Columns of muddy smoke spurt up continually as high explosives tear deeper into this ulcered area. During heavy bombardment and attacks I have seen shells falling like rain. The countless towers of smoke remind one of Gustave Doré's picture of the fiery tombs of the arch-heretics in Dante's "Hell." A smoky pall covers the sector under fire, rising so high that at a height of one thousand feet one is enveloped in its mist-like fumes. Now and then monster projectiles hurtling through the air close by leave one's plane rocking violently in their wake. Aeroplanes have been cut in two by them.

For us the battle passes in silence, the noise of one's motor deadening all other sounds. In the green patches behind the brown belt myriads of tiny flashes tell where the guns are hidden; and those flashes, and the smoke of bursting shells, are all we see of the fighting. It is a weird combination of stillness and havoc, the Verdun conflict viewed from the sky.

Far below us, the observation and range-finding planes circle over the trenches like gliding gulls. At a feeble altitude they follow the attacking infantrymen and flash back wireless reports of the engagement. Only through them can communication be maintained when, under the barrier fire, wires from the front lines are cut. Sometimes it falls to our lot to guard these machines from Germans eager to swoop down on their backs. Sailing about high above a busy flock of them makes one feel like an old mother hen protecting her chicks.

The pilot of an *avion de chasse* must not concern himself with the ground, which to him is useful only for learning his whereabouts. The earth is all-important to the men in the observation, artillery-regulating, and bombardment machines, but the fighting aviator has an entirely different sphere. His domain is the blue heavens, the glistening rolls of clouds below the fleecy banks towering above the vague aerial horizon, and he must watch it as carefully as a navigator watches the storm-tossed sea.

On days when the clouds form almost a solid flooring, one feels very much at sea, and wonders if one is in the navy instead of aviation. The diminutive Nieuports skirt the white expanse like torpedo boats in an arctic sea, and sometimes, far across the cloud-waves, one sights an enemy escadrille, moving as a fleet.

Principally our work consists of keeping German airmen away from our lines, and in attacking them when opportunity offers. We traverse the brown band and enter enemy territory to the accompaniment of an anti-aircraft cannonade. Most of the shots are wild, however, and we pay little attention to them. When the shrapnel comes uncomfortably close, one shifts position slightly to evade the range. One glances up to see if there is another machine higher than one's own. Low, and far within the German lines, are several enemy planes, a dull white in appearance, resembling sandflies against the mottled earth. High above them one glimpses the mosquito-like forms of two Fokkers. Away off to one side white shrapnel puffs are vaguely visible, perhaps directed against a German crossing the lines. We approach the enemy machines ahead, only to find them slanting at a rapid rate into their own country. High above them lurks a protection plane. The man doing the "ceiling work," as it is called, will look after him for us.

Getting started is the hardest part of an attack. Once you have begun diving you're all right. The pilot just ahead turns tail up like a trout dropping back to water, and swoops down in irregular curves and circles. You follow at an angle so steep your feet seem to be holding you back in your seat. Now the black Maltese crosses on the German's wings stand out clearly. You think of him as some sort of a big bug. Then you hear the rapid tut-tut-tut of his machine-gun. The man that dived ahead of you becomes mixed up with the topmost German. He is so close it looks as if he had hit the enemy machine. You hear the staccato barking of his *mitrailleuse* and see him pass from under the German's tail.

The rattle of the gun that is aimed at you leaves you undisturbed. Only when the bullets pierce the wings a few feet off do you become uncomfortable. You see the gunner crouched down behind his weapon, but you aim at where the pilot ought to be—there are two men aboard the German craft—and press on the release hard. Your *mitrailleuse* hammers out a stream of bullets as you pass over and dive, nose down, to get out of range. Then, hopefully, you redress and look back at the foe. He ought to be dropping earthward at several miles a minute. As a matter of fact, however, he is sailing serenely on. They have an annoying habit of doing that, these Boches.

Zeppelins as well as the stationary kite balloons and the swiftly flying aeroplanes often tempted the fighting aviators to attack. One of the most successful of the British champions of the air, though his own life was ended in the second year of the war, was sub-Lieutenant R. A. J. Warneford, of the British Flying Corps. In his brief period of service Warneford won more laurels than any of the British aviators of the time. He was absolutely fearless, with a marvellous control of the fast Vickers scout which he employed, and fertile in every resource of the chase and of the flight. In an interview widely printed at the time, Lieutenant Warneford thus told the story of his casual meeting of a German Zeppelin high in air between Ghent and Brussels and his prompt and systematic destruction of the great balloon. The story as told in his own language reads like the recounting of an everyday event. That to meet an enemy more than a mile above the earth and demolish him was anything extraordinary does not seem to have occurred to the aviator.

> I proceeded on my journey at an increased height. It was just three o'clock in the morning when all of a sudden I perceived on the horizon about midway between Ghent and Brussels a Zeppelin flying fast at an altitude of about six thousand feet. I immediately flew toward it and when I was almost over the monster I descended about fifteen metres, and flung six bombs at it. The sixth struck the envelope of the ship fair and square

in the middle. There was instantly a terrible explosion. The displacement of the air round about me was so great that a tornado seemed to have been produced. My machine tossed upward and then flung absolutely upside down, I was forced to loop the loop in spite of myself. I thought for a moment that the end of everything had come. In the whirl I had the pleasure of seeing my victim falling to the earth in a cloud of flames and smoke. Then by some miracle my machine righted herself and I came to earth in the enemy's country. I was not long on the ground you may be sure. I speedily put myself and my machine into working order again; then I set my engine going.

This time the fortunate aviator returned safely to his own territory. He had then served only four months, had attained the age of twenty-three, and even in so brief a service had received the Cross of the Legion of Honour from France and the Victoria Cross from the British. Only one week after this courageous exploit he was killed while on a pleasure flight and with him a young American journalist, Henry Beach Needham, to whom he was showing the battlefield.

During the early years of the war all of the governments were peculiarly secretive concerning all matters relative to their aviation services. This was probably due to the fact that the flying corps was a brand new branch of the service. No nation was adequately equipped with flyers. Each was afraid to let its enemies know how insufficient were its air guards, or what measures were being taken to bring the aerial fleet up to the necessary point of efficiency. Investigators were frowned upon and the aviators themselves were discouraged from much conversation about their work.

About the beginning of 1916 the British suddenly awoke to the fact that even in war publicity has its value. It was necessary to arouse the enthusiastic support of the people for recruiting or for the conscription which ultimately was ordered. To do this graphic descriptions of what was doing at the front in the various branches of the service seemed necessary. The best writers in England were mobilized for this work. Kipling wrote of the submarines, Conan Doyle of the fighting on the fields of France. The Royal Flying Corps gave out a detailed story the authorship of which was not stated, but which describes most picturesquely the day of a flying man.

In the United States it appeared in the *Sun*, of New York, and sections of it are reprinted here:

"The following bombing will be carried out by No.—Squadron at night (10 P.M., 12 midnight, and 2 A.M.). At each of these times three machines, each carrying eight twenty-pound bombs, will bomb respectively P—, C—, H—."

Thus the operation order read one evening in France. Just an ordinary order too, for bombing is carried out day and night incessantly. Bombing by night is usually carried out on towns and villages known to be resting places of the German troops, and it is part of the work of the Royal Flying Corps to see that the Hun never rests.

Fritz after a hard spell in the trenches is withdrawn to some shell torn village behind his lines to rest. He enters the ruined house, that forms his billet, and with a sigh of contentment at reaching such luxury after the miseries of trench life prepares to sleep in peace. He dreams of home, and then out of the night comes the terror of the air.

A bomb falls in his billet, exploding with a terrific report and doing more damage to the already ruined walls. Possibly a few of his comrades are wounded or killed. Other explosions take place close by and the whole village is in turmoil.

Fritz does not sleep again. His nerves are jangled and all possibility of sleep is gone. The next day he is in a worse condition than after a night in the trenches. This continues night after night. The damage to German morale is enormous.

From the aerial point of view things are different. A pilot warned for night flying takes it as he takes everything else, with apparent unconcern. He realizes that he will have an uninteresting ride in the dark; the danger from "Archie" will be small, for an aeroplane is a difficult target to keep under observation with a searchlight, and the danger from hostile aircraft will be smaller still.

Over the trenches the star shells of the infantry may be seen, occasionally the flash of a badly concealed gun glints in the darkness or the exploding bombs of a trench raiding party cause tiny sparks to glimmer far below. Probably the enemy, hearing the sound of engines, will turn on his searchlights and sweep the sky with long pencils of light. The pilot may be picked up for a second, and a trifle later the angry bang, bang, bang of "Archie" may be heard, firing excitedly at the place where

A French scout aeroplane

the aeroplane ought to be but is not—the pilot has probably dipped and changed his course since he was in the rays of the searchlight. He may be caught again for an instant and the performance is repeated.

Before long the vicinity of the target is reached and he prepares to drop his bombs, usually eight in number. A little before he is over the spot the first bombs will be released, for the trajectory of the bomb follows the course of the machine if the latter keeps on a straight course and when it explodes the aeroplane is still overhead. Down far below will be seen a tiny burst of flame; possibly a large fire blazes up and the pilot knows that his work is good. He then turns and repeats his performance until all his bombs are exhausted, when he turns for home.

Bombs are usually dropped from a low altitude at night in order to be surer of getting the target. If during the performance any local searchlights are turned on "Archie" gets busy and a merry game of hide and seek in and out the beams takes place. If the aeroplane is very low, and bombs are sometimes dropped from a height of only a few hundred feet, it is highly probable that the bursting shells do more damage than the aeroplane's bombs, and it is almost impossible to wing an aeroplane by night.

Over the lines the pilot probably meets more searchlights, dodges them, and gradually descends. Below him he sees the aerodromes of the surrounding squadrons lighted up for landing purposes. Should he be in doubt as to which is his own he fires a certain combination of signal lights and is answered from below. He then lands, hands his machine over to the mechanics, and turns in.

So much for night bombing. By day it is different. Though at night it is the billets which usually form the target, by day bombing is carried out for the purpose of damaging specific objects. Railroads, dumps of stores and ammunition, and enemy aerodromes are the favourite targets.

The raiding machines fly in formation and are surrounded by other machines used solely for protective purposes. Generally a raid is carried out by machines from two squadrons, the bomb carriers belonging to a corps wing and the escorting machines to an army wing.

All the machines meet at a prearranged rendezvous well on our side of the line at a certain time and a given altitude.

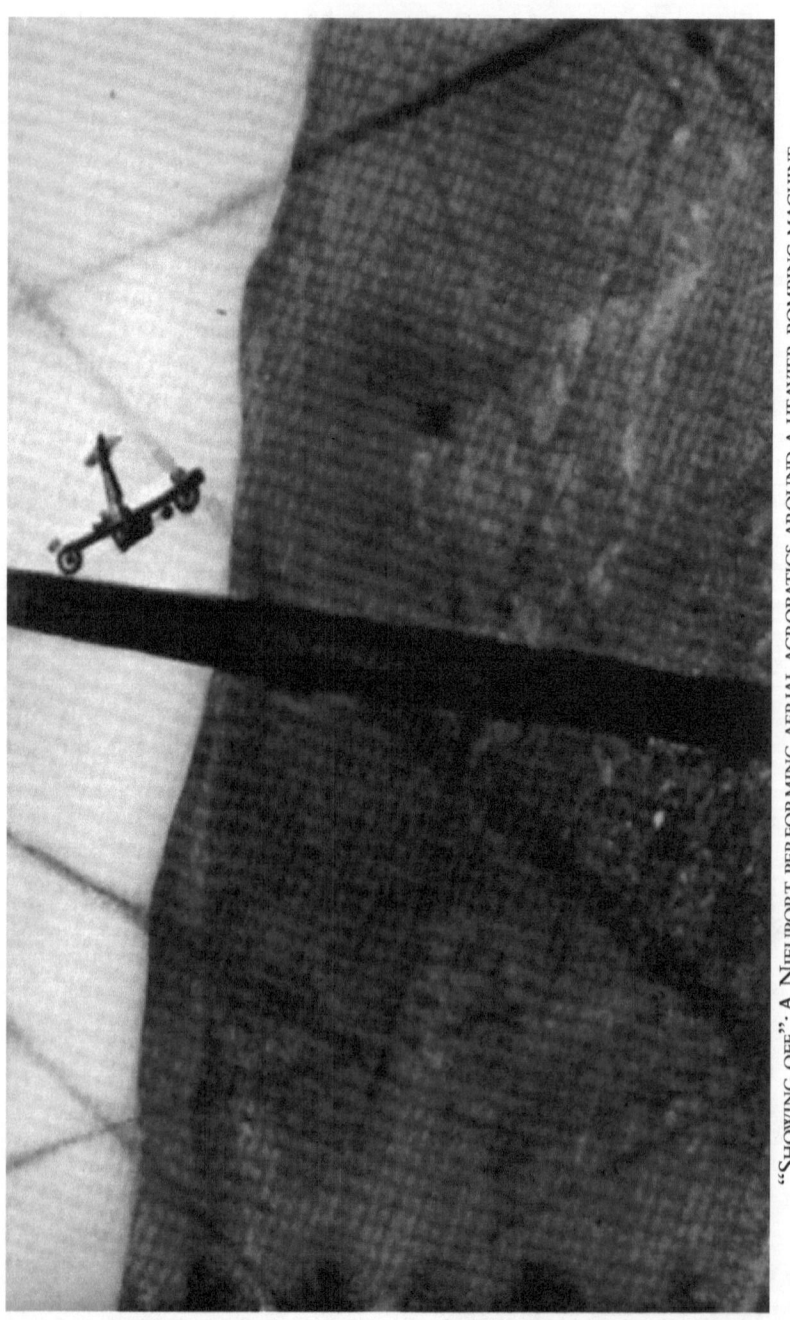

"Showing off": A Nieuport performing aerial acrobatics around a heavier bombing machine

There they manoeuvre into their correct formation. A flight commander leads the raid and his machine is distinguished by streamers tied to it.

Once over the target the fighters scatter and patrol the neighbourhood while the bombers discharge their missiles on the objective. Usually, unless anti-aircraft fire is very heavy, they descend a few thousand feet to make surer of the target, and when their work is completed rise again to the level of the escort.

Results can usually be fairly judged by day. An ammunition dump quickly shows if it is hit and stores soon burst into flame. Railway stations or junctions show clearly damage to buildings or overturned trucks, but the damage to the track itself is hard to estimate. Aerodromes may be bombed for the purpose of destroying enemy machines in their hangars or merely in order to spoil the landing by blowing holes all over the place. It is with great delight that a pilot remarks in his report that a hostile machine, surrounded by mechanics, was about to ascend, but that instead he had descended to within a few hundred feet and obtained a direct hit, with the result that the enemy machine, including the surrounding men, seemed to be severely damaged.

One officer on a bomb raid saw his chance in this way, descended to four hundred feet under intense rifle fire, successfully bombed the enemy machine, which was just emerging from its hangar, and then tried to make off. Unfortunately at this moment his engine petered out, possibly on account of the enemy's fire, and he had to descend.

By skilful planing he managed to descend about three quarters of a mile away, in full view of the enemy. Instead of giving up the ghost and at once firing his machine, this officer jumped out and, utterly unperturbed by the German fire or by the Huns making across country to take him prisoner, commenced to inspect the engine. Luckily he found the cause of the trouble at once, put it right,—it was only a trifling mishap,—adjusted the controls, and swung the propeller.

The engine started, he jumped in, with the nearest Hun only a hundred yards off, and opening the throttle raced over the ground and into the air pursued by a futile fusillade of bullets. His engine held out and he safely regained his aerodrome, after having been reported missing by his comrades. For this escapade he received the Military Cross—a well-earned reward.

When all the bombs have been dropped and the formation resumed the machines head for home. It is on the homeward journey that events may be expected, for time enough has elapsed for the Hun to detail a squadron to intercept our returning machines and pick off any stragglers that may fall behind.

It is a favourite Boche manoeuvre to detail some of his slow machines to entice our fighters away from the main body, and when this has been accomplished, to attack the remainder with Fokkers, which dive from aloft onto the bombing machines. This trick is now well-known and the fighters rarely leave their charges until the latter are in comparative safety.

Sometimes a Hun of more sporting character than his brothers will wait alone for the returning convoy, hiding himself thousands of feet up in the clouds until he sees his moment. Then singling out a machine he will dive at it, pouring out a stream of bullets as he falls. Sometimes he achieves his object and a British machine falls to earth, but whatever the result, the Hun does not alter his tactics. He dives clean through the whole block of machines, down many thousands of feet, only flattening out when close to the ground.

The whole affair is so swift—just one lightning dive—that long before a fighter can reach the Hun the latter is away thousands of feet below and heading for home and safety. Every Fokker pilot knows that once his surprise dive is over he has no chance against another machine—the build of the Fokker only allows this one method of attack—and he does not stop to argue about it. His offensive dive becomes a defensive one—that is the sole difference.

Sometimes a large squadron of German machines, composed of various types of aeroplanes, intercepts a returning formation. If it attacks a grand aerial battle ensues. The British fighting machines spread out in a screen to allow the bombing machines a chance of escape and then attack the Huns as they arrive. In one place one British aeroplane will be defending itself from two or three German machines; close by two or three of our busses will be occupied in sending a Hun to his death; elsewhere more equal combats rage and the whole sky becomes an aerial battlefield, where machines perform marvellous evolutions, putting the best trick flying of pre-war days very much in the shade. No sooner has a pilot accounted for his foe, by killing him, forcing him to descend, or making him think discretion the better part of valour, than he turns to the help of a hard-pressed brother, surprising the enemy by an attack from the rear or otherwise creating a diversion.

A single shot in the petrol tank proves fatal; loss of pressure ensues, the engine fails, and the pilot is forced to descend. He can usually land safely, but should he be in enemy territory he must fire his machine and prepare for a holiday in Germany. Should he be fortunate enough to plane over our lines little damage is done; the tank can be repaired and the machine made serviceable again. But for the time being he is out of the fight. Sometimes the escaping petrol may ignite and the pilot and observer perish in the flames—the most terrible fate of all.

The aerial battle ends in one of two ways: one side is outmanoeuvred, outnumbered, and has lost several machines and flies to safety, or, the more usual ending, both sides exhaust their ammunition, only a limited quantity perforce being carried, and the fight is of necessity broken off. Meanwhile the bombing machines have probably crossed the line in safety, and their duty is finished. Should they be attacked by a stray machine they are armed and quite capable of guarding themselves against any attack except one in force.

During these bomb raids photographs of the target are frequently obtained or should the staff require any district crossed on the journey and taken they are generally secured by bombing machines. It is wonderful what minute details may be seen in a photograph taken at a height of from eight to twelve thousand feet, and our prints, which are far superior to those taken by the Hun, have revealed many useful points which would otherwise have remained unknown.

When it is remembered that a single machine crossing the line is heavily shelled it may be conceived what an immense concentration of "Archies" is made on the raiders on their return. It is remarkable what feeble results are obtained considering the intensity of the bombardment, but rarely is a machine brought down, though casualties naturally occur occasionally.

Lieutenant C., in company with other machines, had successfully bombed his target and had meanwhile been heavily shelled, with the result that his engine was not giving its full number of revolutions and he lagged a little behind the rest of the formation. No hostile aircraft appeared and all went well until he was about to cross the lines, when a terrific bombardment was opened on him.

He dodged and turned to the best of his ability, but a well-aimed shell burst just above him and a piece of the "Archie" hit him on the head, not seriously wounding him, but knocking him unconscious. The machine, deprived of the guiding hand, immediately got into a dive and commenced a rapid descent from ten thousand feet, carrying the unconscious pilot with it, to be dashed to pieces on the ground.

Whether the rush of air, the sudden increase of pressure, or the passing off of the effect of the blow caused the disabled man to come to his senses is not known, but when the machine was only a few hundred feet from the ground, Lieutenant C. recovered his senses sufficiently to realize his position and managed to pull the machine up and make a landing. He then lapsed into unconsciousness again. Had he remained in his state of collapse half a minute longer, he would inevitably have been killed.

Another curious case of wounding was that of Lieutenant H., who was also returning from a bomb raid. When passing through the heavily shelled zone his machine was hit by a shell, which passed through the floor by the pilot's seat and out at the top without exploding. Lieutenant H. thought it must have been very close to his leg, but he was so fully occupied with manoeuvring to dodge other shells that he had no time to think of it.

He crossed the line and began to plane down when he was aware of a feeling of faintness, but pulling himself together he landed his machine, taxied up to the sheds, and attempted to get out. It was only then that he realized that his leg was shot almost completely off above the knee; the lower part was merely hanging by a piece of skin.

Incredible as it may seem the shell which hit his machine also tore through the leg—luckily without exploding—unknown to Lieutenant H. Probably the force of the blow and excitement of the moment caused it to pass unnoticed and the torn nature of the wound helped to close the arteries and prevent his bleeding to death. He recovered, and though no longer flying is still engaged in doing his duty for the duration of the war.

The courage and dash of the American aviators, serving with

Raid on a troop train

the French Army, led the Allies to expect great things of our flying corps which should be organized immediately after our declaration of war. About the time of that declaration Major L. W. B. Rees, of the British Flying Corps, came to the United States for the purpose of giving to our authorities the benefit of British experience in raising and equipping aerial fleets and in the development of the most efficient tactics. Major Rees in an official statement set forth many facts of general interest concerning the various flying services of the belligerent armies. The British, he said, fly on three levels with three different kinds of machines. Nearest the ground, about six thousand feet up, are the artillery directors who hover about cutting big figure eights above the enemy trenches and flash back directions by wireless to the British artillerists. These observers are, of course, exposed to attack from anti-aircraft guns, the effective range of which had by the middle of war become as great as ten thousand feet. Yet, as has already been noted, the amount of execution done by these weapons was surprisingly small. The observers are protected from attack from above, first by the heavy fighting planes, flying at ten thousand feet, carrying two men to the plane and able to keep the air for four hours at a time at a speed of 110 miles an hour. They are supposed to use every possible vigilance to keep the enemy's fighters away from the slower and busy observing machines. In this they are seconded by the lighter one-man fighting machines which cruise about at a height of fifteen thousand feet at a speed of 130 miles an hour and able to make a straight upward dash at the rate of ten thousand feet in ten minutes. The aviators of these latter machines came to describe their task as "ceiling work," suggesting that they operated at the very top of the world's great room. They are able to keep the air only about two hours at a time.

Americans, perhaps, gave exaggerated importance to the work of the Lafayette Escadrille which was manned wholly by American boys, and which, while in service from the very beginning of the war, was the first section of the French Army permitted to display the flag of the United States in battle after our declaration of war. It was made up, in the main, of young Americans of good family and independent means, most of them being college students who had laid down their books for the more exciting life of an airman. They paid heavily in the toll of death for their adventure and for the conviction which led them to take the side of democracy and right in the struggle against autocracy and barbarism months, even years, before their nation finally determined to join with them. In the first two and a half years of the war, seven of the aviators in this comparatively small body lost their lives.

Harvard College was particularly well represented in the American Flying Corps—although this is a proper and pertinent place to say that the sympathy shown for the allied cause by the young collegians of the United States was a magnificent evidence of the lofty righteousness of their convictions and the spirit of democracy with which they looked out upon the world. When the leash was taken off by the declaration of war by the United States the college boys flocked to training camps and enlistment headquarters in a way that bade fair to leave those institutions of learning without students for some years to come.

But to hark back to Harvard, it had in the Lafayette Escadrille five men in 1916; three of these, Kiffen Rockwell, Norman Prince, and Victor Chapman, were killed in that year. A letter published in *Harvard Volunteers in Europe* tells of the way these young gladiators started the day's work:

Rockwell called me up at three: "Fine day, fine day, get up!"

It was very clear. We hung around at Billy's (Lieutenant Thaw) and took chocolate made by his *ordonnance*. Hall and the Lieutenant were guards on the field; but Thaw, Rockwell, and I thought we would take *a tour chez les Boches*. Being the first time the *mechanaux* were not there and the machine gun rolls not ready. However it looked misty in the Vosges, so we were not hurried.

"Rendezvous over the field at a thousand metres," shouted Kiffen.

I nodded, for the motor was turning; and we sped over the field and up.

In my little cockpit from which my shoulders just protrude I have several diversions besides flying. The compass, of course, and the map I keep tucked in a tiny closet over the reservoir before my knees, a small clock and one altimeter. But most important is the contour, showing revolutions of the motor which one is constantly regarding as he moves the *manettes* of gasoline and gas back and forth. To husband one's fuel and tease the motor to round eleven takes attention, for the carburettor changes with the weather and the altitude.... The earth seemed hidden under a fine web such as the Lady of Shalott wove.

A BURNING BALLOON, PHOTOGRAPHED FROM A PARACHUTE BY THE ESCAPING BALLOONIST

Soft purple in the west, changing to shimmering white in the east. Under me on the left the Vosges like rounded sand dunes cushioned up with velvety light and dark masses (really forests), but to the south standing firmly above the purple cloth like icebergs shone the Alps. My! they look steep and jagged. The sharp blue shadows on their western slopes emphasized the effect. One mighty group standing aloof to the west—Mount Blanc perhaps. Ah, there are quantities of worm-eaten fields my friends the trenches—and that town with the canal going through it must be M—. Right beside the capote of my engine, showing through the white cloth a silver snake—the Rhine!

What, not a quarter to six, and I left the field at five! Thirty-two hundred metres. Let's go north and have a look at the map.

While thus engaged a black puff of smoke appeared behind my tail and I had the impression of hearing a piece of iron hiss by. "Must have got my range first shot!" I surmised, and making a steep bank piqued heavily. "There, I have lost them now." The whole art of avoiding shells is to pay no attention till they get your range and then dodge away, change altitude, and generally avoid going in a straight line. In point of fact, I could see bunches of exploding shells up over my right shoulder not a kilometre off. They continued to shell that section for some time; the little balls of smoke thinning out and merging as they crossed the lines.

In the earlier days of the war, when the American aviators were still few, their deeds were widely recounted in their home country, and their deaths were deplored as though a personal loss to many of their countrymen. Later they went faster and were lost in the daily reports. Among those who had early fixed his personality in the minds of those who followed the fortunes of the little band of Americans flying in France was Kiffen Rockwell, mentioned in an earlier paragraph, and one of the first to join the American *escadrille*. Rockwell was in the war from sincere conviction of the righteousness of the Allies' cause.

"I pay my part for Lafayette, and Rochambeau," he said proudly, when asked what he was doing in a French uniform flying for France. And pay he did though not before making the Germans pay heavily for their part. Once, flying alone over Thann, he came upon a German scout. Without hesitation the battle was on. Rockwell's machine was the higher, had the better position. As aerial tactics demanded he dived for the foe, opening fire as soon as he came within thirty or forty yards. At his fourth shot the enemy pilot fell forward in his seat and his machine fell heavily to earth. He lighted behind the German lines much to the victor's disgust, for it was counted a higher achievement to bring your foe to earth in your own territory. But Rockwell was able to pursue his victim far enough to see the wreck burst into flames.

Though often wounded, Rockwell scorned danger. He would go into action so bandaged that he seemed fitter to go to an hospital. He was always on the attack—"shoved his gun into the enemy's face" as his fellows in the *escadrille* expressed it. So in September, 1916, he went out after a big German machine, he saw flying in French territory. He had but little difficulty in climbing above it, and then dashed down in his usual impetuous manner, his machine gun blazing as he came on. But the German was of heavier metal mounting two machine guns. Just as to onlookers it seemed that the two machines would crash together, the wings of one side of Rockwell's plane suddenly collapsed and he fell like a stone between the lines. The Germans turned their guns on the pile of wreckage where he lay, but French gunners ran out and brought his body in. His breast was all blown to pieces with an explosive bullet—criminal, of course, barbarous and uncivilized, but an everyday practice of the Germans.

Rockwell was given an impressive funeral. All the British pilots, and five hundred of their men marched, and the bier was followed by a battalion of French troops. Over and around the little French graveyard aviators flew dropping flowers. In later days less ceremony attended the last scene of an American aviator's career.

Another American aviator, also a Harvard man, who met death in the air, was Victor Chapman of New York, a youth of unusual charm, high ideals, and indomitable courage. At the very outbreak of the war he enlisted in the French Foreign Legion—a rough entourage for a college-bred man. Into the Foreign Legion drifted everything that was doubtful, and many that were criminal. No questions were asked of those who sought its hospitable ranks, and readers of Ouida's novel *Under Two Flags* will recall that it enveloped in its convenient obscurity British lordlings and the lowest of Catalonian thieves. But in time of actual war its personnel was less mixed, and Chapman's letters showed him serving there contentedly as pointer of a *mitrailleuse*. But not for long. Most of the spirited young Americans who entered the French Army aspired to serve in the aviation corps, and Chapman soon was transferred to that field. There he developed into a most daring flyer. On one occasion, with a bad scalp wound, after a brush with four German machines, he made his landing with his machine so badly wrecked that he had to hold together the broken ends of a severed control with one hand, while he steered with the other. Instead of laying up for the day he had his mechanician repair his machine while a surgeon repaired him, then, patched up together, man and machine took the air again in search for the Boches.

In June, 1916, though still suffering from a wound in the head, he started in his machine to carry some oranges to a comrade lying desperately wounded in a hospital some miles away. On the way he saw in the distance behind the German lines two French airmen set upon by an overwhelming force of Germans. Instantly he was off to the assistance of his friends, plunging into so unequal a fight that even his coming left the other Americans outnumbered. But he had scarce a chance to strike a blow. Some chance shot from a German gun put him out of action. All that the other two Americans, Lufbery and Prince, knew was that they saw a French machine come flying to their aid, and suddenly tip and fall away to earth. Until nightfall came and Chapman failed to return none was sure that he was the victim.

The part played by young Americans as volunteers for France before the United States entered upon the war was gallant and stimulating to national pride. It showed to the world—and to our own countrymen who needed the lesson as much as any—that we had among our youth scores who, moved by high ideals, stood ready to risk their lives for a sentiment—stood ready to brave the myriad discomforts of the trenches, the bursting shrapnel, the mutilating liquid fire, the torturing gas that German autocracy should be balked of its purpose of dominating the world.

And the service of these boys aided far more than they knew. The fact that our countrymen in numbers were flying for France kept ever before the American people the vision of that war in the air of which poets and philosophers had dreamed for ages. It brought home to our people the importance of aviation before our statesmen could begin to see it. It set our boys to reading of aircraft, building model planes, haunting the few aviation fields which at the time our country possessed. And it finally so filled the consciousness of our people with conviction of the supreme importance of aviation as an arm of the national armed service that long before the declaration of war the government was embarrassed by the flood of volunteers seeking to be enrolled in the flying forces of the nation.

CHAPTER 9

The United States at War

The entrance of the United States upon the war was the signal for a most active agitation of the question of overwhelming the enemy with illimitable fleets of aircraft. Though the agitation was most vociferous in this country whence it was hoped the enormous new fleets of aircraft would come, it was fomented and earnestly pressed by our Allies. France sent a deputation of her leading flyers over to supervise the instruction of our new pilots. England contributed experts to advise as to the construction of our machines. The most comprehensive plans were urged upon Congress and the Administration for the creation of a navy of the air. A bill for an initial appropriation of $640,000,000, for aircraft purposes alone, was passed and one for a Department of Aeronautics to be established, co-ordinate with those of War and the Navy, its secretary holding a seat in the cabinet, was introduced in Congress. Many of the most eminent retired officers of the navy joined in their support. Retired officers only because officers in active service were estopped from political agitation.

There was every possible reason for this great interest in the United States in wartime aviation. The nation had long been shamefaced because the development of the heavier-than-air machines, having their origin undoubtedly in the inventive genius of Professor Langley and the Wrights, had been taken away from us by the more alert governments of France and Germany. The people were ready to buy back something of our lost prestige by building the greatest of air fleets at the moment when it should exercise the most determinative influence upon the war.

But more. We entered upon the war in our chronic state of unpreparedness. We were without an army and without equipment for one. To raise, equip, and drill an army of a million, the least number that would have any appreciable effect upon the outcome of the war, would take months. When completed we would have added only to the numerical superiority of the Allies on the Western Front. The quality of a novel and decisive contribution to the war would be lacking.

So too it was with our navy. The British Navy was amply adequate to deal with the German fleet should the latter ever leave its prudent retreat behind Helgoland and in the bases of Kiel and Wilhelmshaven. True it was not capable of crushing out altogether the submarine menace, but it did hold the German underwater boats down to a fixed average of ships destroyed, which was far less than half of what the Germans had anticipated. In this work our ships, especially our destroyers, took a notable part.

The argument for a monster fleet of fighting aircraft, thus came to the people of the United States in a moment of depression and perplexity. By land the Germans had dug themselves in, holding all of Belgium and the thousands of square miles of France they had won in their first dash to the Marne. What they had won swiftly and cheaply could only be regained slowly and at heavy cost. True, the Allies were, day by day, driving them back from their position, but the cost was disheartening and the progress but slow.

By sea the Germans refused to bring their fleet to battle with their foes. But from every harbour of Belgium, and from Wilhelmshaven and Kiel, they sent out their sinister submarines to prey upon the commerce of the world—neutral as well as belligerent. Against them the navies of the world were impotent. To the threat that by them Germany would starve England into cowering surrender, the only answer was the despairing effort to build new ships faster than the submarines could sink those afloat—even though half a million tons a month were sent to the bottom in wasteful destruction.

Faced by these disheartening conditions, wondering what they might do that could be done quickly and aid materially in bringing the war to a triumphant conclusion, the American people listened eagerly to the appeals and arguments of the advocates of a monster aerial fleet.

> Listen (said these advocates), we show you a way to spring full panoplied into the war, and to make your force felt with your first stroke. We are not preaching dreadnoughts that take four years to build. We are not asking for a million men taking nearly a year to gather, equip, drill, and transport to France, in imminent danger of destruction by the enemy's submarines every mile of the way.

A Caproni biplane circling the Woolworth Building

We ask you for a cheap, simple device of wood, wire, and cloth, with an engine to drive it. All its parts are standardized. In a few weeks the nation can be equipped to turn out 2000 of them weekly. We want within the year 100,000 of them. We do not ask for a million men. We want 10,000 bright, active, hardy, plucky American boys between 20 and 25 years of age. We want to give them four months' intensive training before sending them into the air above the enemy's lines. In time we shall want 25,000 to 35,000 but the smaller number will well do to open the campaign.

And what will they effect?

Do you know that to-day the eyes of an army are its aeroplanes? Cavalry has disappeared practically. If a general wishes to pick out a weak point in his enemy's line to assault he sends out airmen to find it. If he is annoyed by the fire of some distant unseen battery over the hills and far away he sends a man in an aeroplane who brings back its location, its distance, and perhaps a photograph of it in action. If he suspects that his foe is abandoning his trenches, or getting ready for an attack, the ready airmen bring in the facts.

And of course the enemy's airmen serve their side in the same manner. They spy out what their foe is doing, and so far as their power permits prevent him from seeing what they are doing.

Now suppose one side has an enormous preponderance of aircraft—six to one, let us say. It is not believed, for example, that at this moment Germany has more than 10,000 aircraft on the whole western front. Let us imagine that through the enterprise of the United States our Allies were provided with 25,000 on one sector which we intended to make the scene of an attack on the foe. Say the neighbourhood of Arras and Lille. For days, weeks perhaps, we would be drawing troops toward this sector from every part of the line. Through the reports of spies the enemy's suspicions would be aroused. It is the business of an efficient general to be suspicious. He would send out his aeroplanes to report on the activities of the other side. Few would come back. None would bring a useful report. For every German plane that showed above the lines three Allied planes would be ready to attack and destroy it or beat it back. The air would be full of Allied airmen—the great bombing planes flying low and inundating the trenches with bombs, and the

Cruising at 2000 feet—one biplane photographed from another

troops on march with the deadly *fléchette*s. Over every German battery would soar the observation plane indicating by tinsel or smoke bombs the location of the guns, or even telegraphing it back by wireless to the Allied batteries safe in positions which the blinded enemy could never hope to find. Above all in myriads would be soaring the swift fighting scouts, the Bleriots, Nieuports, Moranes or perhaps some new American machine to-day unknown. Let the wing of a Boche but show above the smoke and they would be upon him in hordes, beating him to the ground, enveloping him in flames, annihilating him before he had a chance to observe, much less to report.

What think you would be the result on that sector of the battle line? Why the foe would be cut to pieces, demolished, obliterated. Blinded, he would be unrelentingly punished by an adversary all eyes. Writhing under the concentrated fire of a thousand guns he could make no response, for his own guns could not find the attacking batteries. Did he think to flee? His retreating columns would be marked down by the relentless scouts in the air, and the deadly curtain of fire from well-coached batteries miles away would sweep every road with death. If in desperation he sought to attack he would do so ignorant whether he were not hurling his regiments against the strongest part of the Allied line, and with full knowledge of the fact that though he was blinded they had complete information of his strength and dispositions.

The argument impressed itself strongly upon the mind of the country. There appeared indeed no public sentiment hostile to it nor any organized opposition to the proposition for an enormous appropriation for purposes of aviation. The customary inertia of Congress delayed the actual appropriation for some months. But the President espoused its cause and the Secretaries both of War and the Navy warmly recommended it, although they united in opposing the proposition to establish a distinct department of aeronautics with a seat in the Cabinet. Being human neither one desired to let his share of this great new gift of power slip out of his hands. Leading in the fight for this legislation was Rear-Admiral Robert E. Peary, U. S. N., retired, the discoverer of the North Pole. Admiral Peary from the very outbreak of the war consecrated his time and his abilities to pushing the development of aeronautics in the United States. He was continually before Congressional committees urging the fullest appropriations for this purpose. In his first statement before the Senate Committee he declared that "in the immediate future the air service will be more important than the army and navy combined," and supported that statement by reference to utterances made by such British authorities as Mr. Balfour, Lord Charles Beresford, Lord Northcliffe, and Lord Montague. In an article published shortly after his appearance before the Senate Committee, the Admiral summarized in a popular way his views as to the possibility of meeting the submarine menace with aircraft, and what the United States might do in that respect. He wrote:

> We are receiving agreeable reports as to the efficiency of the American destroyer flotilla now operating against submarines in the North Sea. An unknown naval officer, according to the newspapers of May 30th, calls for the immediate construction of from 100 to 200 additional American destroyers.
>
> By all means let us have this force—when it can be made ready—but it would take at least two years to construct, equip, and deliver such a heavy additional naval tonnage, while 200 fighting seaplanes, with a full complement of machine guns, bombs, microphones, and aerial cameras, could be put in active service in the North Sea within six months.
>
> Seaplanes, small dirigibles on the order of the English "blimp" type, and kite balloons have already shown themselves to be more effective in detecting submarines than are submarine chasers or armed liners.
>
> Not only have the British, French, German, and Turkish forces destroyed trawlers, patrol boats, and transports by aircraft, but successful experiments in aeroplane submarine hunting have also been made in this country.

In September, 1916, our first Aerial Coast Patrol Unit, in acting as an auxiliary to the Mosquito Squadron in the annual manoeuvres of the Atlantic fleet, detected objects smaller than the latest type of German submarines from fifteen to twenty feet below the surface.

A more complete aerial submarine hunt took place on March 26th of this year. This was the real thing, because the fliers were looking for German U-boats. Inasmuch as the Navy Department is still waiting before establishing its first and only aeronautical base on the Atlantic seaboard, the honour of having conducted the first aerial hunt of the enemy submarines in American history went to the civilian aviators who are soon to be a part of the Aerial Reserve Squadron at Governor's Island and to the civilian instructors and aerial reservists connected with the Army Aviation School at Mineola, Long Island.

These hawks of the air darted up and down the coast in search of the enemy, often flying as far as eleven miles out to sea. The inlets and bays were searched, vessels plotted, compass direction and time when located were given.

No enemy submarines were found. It developed that the supposed submarines were two patrol motor-boats returning from a trial trip. Nevertheless the incident is illuminating, and the official statement of the Navy Department closed with the words: "This incident emphasizes the need of hydro-aeroplanes for naval scouting purposes."

It is also interesting to note what happened when Lawrence Sperry went out to sea one day last summer in his hydroplane and failed to return. Two seaplanes and three naval destroyers were sent in search of him. In forty minutes the seaplanes returned with the news that they had located Sperry floating safely on the water. At the end of the day, after several hours of search, the destroyers came back without having seen Sperry at all.

Those who may still believe that we Americans cannot build aircraft and that all the exploits we read so much about in the newspapers taking place on the other side are being done in foreign aircraft will be surprised to know that a large number of the big flying boats now in use in the English navy, harbour, and coast defence work are Curtiss machines, designed and built in this country by Americans, with American material and American engines. Great Britain wants all the machines of this type that it can get, and sees no reason why we cannot do the same thing in protecting our own Atlantic seaboard. I quote from C. G. Grey, editor of *The London Aeroplane*:

> Curiously enough, these big flying boats originated in America, and, if America is seriously perturbed about the fate of American shipping and American citizens travelling by sea in the vicinity of Europe, it should not be a difficult matter for America to rig up in a very small space of time quite a fleet of seaplane carriers suitable for the handling of these big seaplanes. If each seaplane ship were armed with guns having a range of five to ten miles, and if the gunners were practised in co-operating with aeroplane spotters, such ships ought to be the very best possible insurance for American lives and goods on the high seas.

I quote from The Associated Press report from Paris on May 14th to show the relative importance of aeroplanes in submarine attacks:

During the last three months French patrol boats have had twelve engagements with submarines, French hydro-aeroplanes have fought them thirteen times, and there have been sixteen engagements between armed merchantmen and submarines.

Henry Woodhouse, one of the most distinguished authorities on aeronautics in the United States, in his standard *Textbook on Naval Aeronautics*, published by the Century Company, has assembled the following data on submarine and aeroplane combats:

On May 4, 1915, the German Admiralty reported an engagement between a German dirigible and several British submarines in the North Sea. The submarines fired on the dirigible without success, whereas bombs from the dirigible sank one submarine.

On May 31, 1915, the German Admiralty announced the sinking of a Russian submarine by bombs dropped by German naval aviators near Gotland.

On July 1, 1915, the Austrian submarine *U-11* was destroyed in the Adriatic by a French aeroplane, which swooped suddenly and dropped three bombs directly on the deck of the submarine. The craft was destroyed and the entire crew of twenty-five were lost.

On July 27, 1915, a German submarine in the Dardanelles was about to launch a torpedo at a British transport filled with troops and ammunition, when British aviators gave the alarm to the transport, and immediately began dropping bombs at the submarine, which had to submerge and escape hurriedly, without launching its torpedo.

On August 19, 1915, the Turkish War Office stated that an Allied submarine had been sunk in the Dardanelles by a Turkish aeroplane.

On August 26, the Secretary of the British Admiralty announced that Squadron Commander Arthur W. Bigsworth in a single-handed attack bombed and destroyed a German submarine off Ostend.

Lieutenant Viney received the Victoria Cross and Lieutenant de Sincay was recommended for the Legion of Honour for having flown over a German submarine and destroyed it with bombs off the Belgian coast on November 18, 1915.

Early in 1916 an Austrian seaplane sank the French submarine *Foucault* in the southern Adriatic. Lieutenant Calezeny was the pilot and the observer was Lieutenant von Klinburg. After crippling the submarine they then performed the remarkable feat of calling another Austrian seaplane and rescuing the entire French crew, two officers and twenty seven men, in spite of the fact that a high sea was running at the time.

It will be noted that Admiral Peary lays great stress on the supreme value of aircraft as foes of the submarine. This was due to the fact that at about the time of his appearance before the Senate Committee the world was fairly panic-stricken by the vigour and effect of the German submarine campaign and its possible bearing upon the outcome of the war. Of that campaign I shall have more to say in the section of this book dealing with submarines. But the subject of the undersea boat in war became at this time inextricably interwoven with that of the aerial fleets, and the sudden development of the latter, together with the marked interest taken in it by our people, cannot be understood without some description of the way in which the two became related.

From the very beginning of the war the Germans had prosecuted a desultory submarine warfare on the shipping of Great Britain and had extended it gradually until neutral shipping also was largely involved. All the established principles of international law, or principles that had been supposed to be established, were set at naught. In bygone days enemy merchant ships were subject to destruction only after their crews had been given an opportunity to take to the boats. Neutral ships bearing neutral goods, even if bound to an enemy port, were liable to destruction only if found upon visit to be carrying goods that were contraband of war. The list of contraband had been from time immemorial rigidly limited, and confined almost wholly to munitions of war, or to raw material used in their construction. But international law went by the board early in the war. Each belligerent was able to ascribe plausible reasons for its amendment out of recognizable form. Great Britain established blockades two hundred miles away from the blockaded ports because the submarines made the old practice of watching at the entrance of the port too perilous. The list of contraband of war was extended by both belligerents until it comprehended almost every useful article grown, mined, or manufactured. But the amendment to international law which acted as new fuel for the flames of war, which aroused the utmost world-wide indignation,

An air battle in progress

and which finally dragged the United States into the conflict, was that by which Germany sought to relieve her submarine commanders of the duty of visiting and searching a vessel, or of giving its people time to provide for their safety, before sinking it.

The German argument was that the submarine was unknown when the code of international law then in force was formulated. It was a peculiarly delicate naval weapon. Its strength lay in its ability to keep itself concealed while delivering its attack. If exposed on the surface a shot from a small calibred gun striking in a vital point would instantly send it to the bottom. If rammed it was lost. Should a submarine rise to the surface, send an officer aboard a ship it had halted, and await the result of his search, it would be exposed all the time to destruction at the hands of enemy vessels coming up to her aid. Indeed if the merchantman happened to carry one gun a single shot might put the assailant out of business. Accordingly the practice grew up among the Germans of launching their torpedoes without a word of warning at their helpless victim. The wound inflicted by a torpedo is such that the ship will go down in but a few minutes carrying with it most of the people aboard. The most glaring, inexcusable, and criminal instance of this sort of warfare was the sinking without warning of the great passenger liner, *Lusitania*, by which more than eleven hundred people were drowned, one hundred and fourteen of them American citizens.

Against this policy—or piracy—the United States protested, and people of this country waxed very weary as month after month through the years 1915 and 1916 Germany met the protests with polite letters of evasion and excuse continuing the while the very practice complained of. But late in January, 1917, her government announced that there would be no longer any pretence of complying with international law, but that with the coming month a campaign of unlimited submarine ruthlessness would be begun and ships sunk without warning and irrespective of their nationality if they appeared in certain prohibited zones. Within twenty-four hours the United States sent the German Ambassador from the country and within two months we were at war.

At once the submarine was seen to be the great problem confronting us. Its attack was not so much upon the United States, for we are a self-contained nation able to raise all that we need within our own borders for our own support. But England is a nation that has to be fed from without. Seldom are her stores of food great enough to avert

A Curtis hydro–aeroplane

starvation for more than six weeks should the steady flow of supply ships from America and Australia to her ports be interrupted. This interruption the Germans proposed to effect by means of their underwater boats. Von Tirpitz and other leaders in the German administration promised the people that within six weeks England would be starved and begging for peace at any price. The output of submarines from German navy yards was greatly increased. Their activity became terrifying. The Germans estimated that if they could sink 1,000,000 tons of shipping monthly they would put England out of action in two or three months. For some weeks the destruction accomplished by their boats narrowly approached this estimate, but gradually fell off. At the same time there was no period in 1917 up to the time of Admiral Peary's statement, or indeed up to that of the preparation of this book, when it was not felt that the cause of the Allies was in danger because of the swarms of German submarines.

It was that feeling, coupled with the wide-spread belief that aircraft furnished the best means of combating the submarine, that caused an irresistible demand in the United States for the construction of colossal fleets of these flying crafts. Congress enacted in midsummer the law appropriating $640,000,000 for the construction of aircraft and the maintenance of the aerial service. The Secretaries of War and the Navy each appealed for heavy additional appropriations for aerial service. The arguments which have already been set forth as supporting the use of aircraft in military service were paralleled by those who urge its unlimited use in naval service.

> Consider (said they) the primary need for attacking these vipers of the sea in their nests. Once out on the broad Atlantic their chances of roaming about undetected by destroyers or other patrol boats are almost unlimited. But we know where they come from, from Kiel, Antwerp, Wilhelmshaven, Ostend, and Zeebrugge. Catch them there and you will destroy them as boys destroy hornets by smoking out their nests. But against this the Germans have provided by blocking every avenue of approach save one. The channels are obstructed and mined, and guarded from the shore by heavy batteries. No hostile ships dare run that gauntlet. Even the much-boasted British navy in the three years of the war has not ventured to attack a single naval base. You could not even seek out the submarines thus sheltered by other submarines because running below the sur-

face our boats could not detect either mines or nets and would be doomed to destruction. The enemy boats come out on the surface protected by the batteries and naval craft. But the air cannot be blocked by any fixed defences. Give us more and more powerful aircraft than the Germans possess and we will darken the sky above the German bases with the wings of our aeroplanes, and rain explosive shells upon the submarines that have taken shelter there until none survive.

The one essential is that our flyers shall be in overwhelming numbers. We must be able not only to take care of any flying force that the Germans may send against us, but also to have enough of our aircraft not engaged in the aerial battle to devote their entire attention to the destruction of the enemy forces below.

From every country allied with us came approval of this policy. At the time the debate was pending in Congress our Allies one after another were sending to us official commissions to consult upon the conduct of the war, to give us the benefit of their long and bitter experience in it, and to assist in any way our preparations for taking a decisive part in that combat. The subject of the part to be played by aircraft was one frequently discussed with them. With the French commission came two members of the staff of General Joffre, Major Tulasne and Lieutenant de la Grange, experts in aviation service. A formal interview given out by these gentlemen expressed so clearly the point of view on aviation and its possibilities held in France where it has reached its highest development that some extracts from it will be of interest here:

"At the beginning of the war the Germans were the only ones who had realized the great importance of aviation from a military point of view," said these officers.

"France had looked upon aviation as a sport, Germany as a powerful weapon in war. This is illustrated by the fact that even in August, 1914, German artillery fire was directed by aeroplanes.

"It was only after the retreat from Belgium and the battle of the Marne that the Allies realized the great importance of aviation. Between August 15 and 25 the French General Staff thought that the greater part of the German army was concentrated in Alsace and that only a few army corps were coming through Belgium. It was only through the reports of the aviators that they realized that this was a mistake and that almost the whole of the German army was invading Belgium.

"Immediately after the battle of the Marne the greatest efforts were made in France to develop the aviation corps in every possible way. The English army, then in process of formation, profited by the experience of the French. Since that time the allied as well as the German aviation corps has grown constantly.

"A modern army is incomplete if it has not a strong aviation corps. All the different services are obliged to turn to the aviation corps for help in their work. An army without aeroplanes is like a soldier without eyes. An army which has the superiority in aviation over its adversary will have the following advantages:

"It will have constantly the latest information on the movements of the enemy. In this way, no concentration of troops will be ignored and no surprise attack will be possible. The attack against the enemy positions will be rendered easier because all the details of these positions will be thoroughly known beforehand. The artillery fire will be much more accurate. Many enemy machines will be brought down by the superior fighting machines and the result will be to strengthen the morale both of the aviators and of the army."

The next question put to the French experts was: "Why do we need to make a great effort to obtain the superiority in the air?" They answered with much interesting detail:

"Because the Germans have understood the importance of aviation from a military point of view and have concentrated all their forces to develop this service.

"Owing to the large number of scientists and technicians they possess they are able constantly to perfect motors and planes. Owing to their great industrial organization they are able to produce an enormous number of the best machines.

"The German aviation service is now fully as strong as that of the Allies as far as numbers are concerned. The superiority in the air can only remain in the hands of the Allies because of the spirit of self-sacrifice of their aviators and their greater skill.

"Germany feels that the decisive phase of the war is imminent and the efforts she will make next year will be infinitely greater than any she has made before. She will try in every way to regain the supremacy of the air. Realizing what a formidable enemy America can be in the air, she will strengthen her aviation forces in consequence.

"The aeroplane is by far the most powerful of all the modern weapons. If the Allies have the supremacy of the air the German artillery will lose its accuracy of aim. It is impossible, because of the long range, for modern guns to fire without the help of aeroplanes. The accuracy of artillery fire depends entirely on its being directed by an aeroplane.

"This was clearly illustrated during the battle of the Somme in 1916. The French at that time had concentrated such a large number of fighting machines that no German machine was allowed to fly over the lines. On the other hand, the Allies' reconnaissance machines were so numerous that each French battery could have its fire directed by an aeroplane.

"The destruction of the enemy positions was in consequence carried out very effectively and very rapidly, while the Germans were obliged to fire blindly and scatter their shells over large areas, incapable as they were of locating our battery emplacements and the positions of our troops. Unluckily, a few weeks later the Germans had called from the different parts of the line a good many of their squadrons, and were able to carry out their work under better conditions.

"We need such a superiority that it will be impossible for any German aeroplane to fly anywhere near the lines.

"Every German kite balloon, every aeroplane would immediately be attacked by a number of allied machines. In this way the German aviation will not only be dominated but will be entirely crushed.

"If we can prevent the Germans from seeing, through their aeroplanes, what we are preparing we will be very near the end of the war. It will require a huge effort to carry out this plan. Neither the English nor the French are able to do so by their own means.

"As far as France is concerned, she is able to keep on building machines rapidly enough to increase her aviation corps at about the same rate as Germany is increasing hers. If she wanted to double or triple her production of machines she could do so, but she would have to call back from the trenches a certain number of skilled workmen, and this would weaken her fighting power. She needs in the trenches all the men who are able to carry a rifle.

"If the Allies are to have the absolute supremacy of the air which we have been describing it will be the privilege of America to give it to them. We want three or four or even five allied machines for one German. America only has the possibilities of production which would allow her to build an enormous number of machines in a very short time.

"The aeroplane is a great engine of destruction. It tells the artillery where to fire, it drops bombs, it gives the enemy all the information he needs to plan murderous attacks. Drive the German aeroplanes down and you will save the lives of thousands of men in our trenches. As Ulysses in the cavern put out the eye of the Cyclops, so the eyes of the beast must be put out before you can attempt to kill it."

Major Tulasne and Lieutenant de la Grange then outlined what the aviation programme of the United States should be, saying:

"American industry must be enabled to begin building at once. No time must be lost in experiments. America must profit by the experience of the Allies. She must choose the best planes and build thousands of them.

"She must build reconnaissance machines which she will need for her army; she must build a large number of fighting machines because it is these machines that will destroy German planes; she must also build squadrons of powerful bombing machines which will go behind the German lines to destroy the railway junctions and bomb the enemy cantonments, so as to give the soldiers no rest even when they have left the trenches.

"Bombing done by a few machines gives poor results. The same cannot be said of this operation carried out by a large number of machines which can go to the same places and bomb continually.

"Besides the number of men that are actually killed in these raids, great disturbance is caused in the enemy's communication lines, thereby hindering the operations. For example, since the British Admiralty has increased the number of its bombing squadrons in northern France and has decided to attack constantly the two harbours of Ostend and Zeebrugge and the locks, bridges, and canals leading to them they have greatly interfered with the activity of these two German bases.

"It is certain that shortly, owing to this, these two ports will no more be used by German torpedo boats and submarines. What the English Royal Naval Air Service has been able to accomplish with 100 machines the Flying Corps of the United States with 1000 machines must be able to carry out on other parts of the front.

"The work of the bombing machines is rendered difficult now by the fact that the actual lines are far from Germany. But it is hoped that soon fighting will be carried on near the enemy frontier and then a wonderful field will be opened to the bombing machines.

"All the big ammunition factories which are in the Rhine and Ruhr valleys, like Krupp's, will be wonderful targets for the American bombing machines. If these machines are of the proper type—that is to say, sufficiently fast and well armed and able to carry a great weight of bombs—nothing will prevent them from destroying any of these important factories.

"As Germany at the present time is only able to continue the war because of her great stock of war material the destruction of her sources of production would be the end of her resistance. For this also the Allies must turn to America. Such a large number of machines is required to produce results that America must be relied on to manufacture them.

"Every man in this country must know that it is in the power of the United States, no matter what can be done in other fields, to bring the war to an end simply by concentrating all its energies on producing an enormous amount of material for aviation, and to enlist a corresponding number of pilots. But this will not be done without great effort. In order to be ready for the great 1918 offensive work must be begun at once."

The extreme secrecy which in this war has characterized the operation of the governments—our own most of all—makes it impossible to state the amount of progress made in 1917 in the construction of our aerial fleet. During the debate in Congress orators were very outspoken in their prophecies that we should outnumber the Kaiser's flying fleet two or three to one. The press of the nation was so very explicit in its descriptions of the way in which we were to blind the Germans and drive them from the air that it is no wonder the Kaiser's government took alarm, and set about building additional aircraft with feverish zeal. In this it was imitated by France and England. It seemed, all at once about the middle of 1917, that the whole belligerent world suddenly recognized the air as the final battlefield and began preparations for its conquest.

All statistical estimates in war time are subject to doubt as to their accuracy—and particularly those having to do in any way with the activities of an enemy country. But competent estimators—or at any rate shrewd guessers—think that Germany's facilities for constructing aeroplanes equal those of France and England together. If then all three nations build to the very limit of their abilities there will be a tie, which the contribution of aircraft from the United States will settle overwhelmingly in favour of the Allies. How great that contribution may be cannot be foretold with certainty at this moment. The building of aircraft was a decidedly infant industry in this country when war began. In the eight years prior to 1916 the government had given orders for just fifty-nine aircraft—scarcely enough to justify manufacturers in keeping their shops open. Orders from foreign governments, however, stimulated production after the war began so that when the United States belatedly took her place as national honour and national safety demanded among the Entente Allies, Mr. Howard E. Coffin, Chairman of the Aircraft Section of the Council of National Defence was able to report eight companies capable of turning out about 14,000 machines in six months—a better showing than British manufacturers could have made when Great Britain, first entered the war.

A feature in the situation which impressed both Congress and the American people was the exposure by various military experts of the defenceless condition of New York City against an air raid by a hostile foreign power. At the moment, of course, there was no danger. The only hostile foreign power with any considerable naval or aerial force was Germany and her fleet was securely bottled up in her own harbours by the overpowering fleet of Great Britain. Yet if one could imagine the British fleet reduced to inefficiency, let us say by a futile, suicidal attack upon Kiel or Heligoland which would leave it crippled, and free the Germans, or if we could conceive that the German threat to reduce Great Britain to subjection by the submarine campaign, proved effective, the peril of New York would then be very real and very immediate. For, although the harbour defences are declared by military authorities to be practically impregnable against attack by sea, they would not be effective against an attack from the air. A hostile fleet carrying a number of seaplanes could round-to out of range of our shore batteries and loose their flyers who could within less than an hour be dropping bombs on the most congested section of Manhattan Island. It is true that our own navy would have to be evaded in such case, but the attack might be made from points more distant from New York and at which no scouts would ever dream of looking for an enemy.

The development in later months of the big heavily armed cruising machines makes the menace to any seaport city like New York still greater. The Germans have built great biplanes with two fuselages, or bodies, armoured, carrying two machine guns and one automatic rifle to each body. They have twin engines of three hundred and forty horse power and carry a crew of six men. They are able in an emergency to keep the air for not less than three days. It is obvious that a small fleet of such machines launched from the deck of a hostile squadron, let us say in the neighbourhood of Block Island, could menace equally Boston or New York, or by flying up the Sound could work ruin and desolation upon all the defenceless cities bordering that body of water.

Nor are the Germans alone in possessing machines of this type. The giant Sikorsky machines of Russia, mentioned in an earlier chapter, have during the war been developed into types capable of carrying crews of twenty-five men with guns and ammunition. The French, after having brought down one of the big German machines with the double bodies, instantly began building aircraft of their own of an even superior type. Some of these are driven by four motors and carry eleven persons, besides guns and ammunition. The Caproni machines of Italy are even bigger—capable of carrying nine guns and thirty-five men. The Congressional Committee was much impressed by consideration of what might be done by a small fleet of aircraft of this type launched from a hostile squadron off the Capes of Chesapeake Bay and operating against Washington. It is not likely that any foreign foe advancing by land could repeat the exploit of the British who burned the capitol in 1812. But in our present defenceless state a dozen aircraft of the largest type might reduce the national capitol to ruins.

If an enemy well provided with aerial force possesses such power of offence an equal power of defence is given to the nation at all well provided with flying craft. In imitation, or perhaps rather in modification, of the English plan for guarding the coasts of Great Britain, a well matured system of defending the American coasts has been worked out and submitted to the national authorities. It involves the division of the coasts of the United States into thirteen aeronautical districts, each with aeronautical stations established at suitable points and all in communication with each other. Eight of these districts would be laid out on the Atlantic Coast extending from the northern boundary of Maine to the Rio Grande River.

Just what the purpose and value of these districts would be may be explained by taking the case, not of a typical one, but of the most important one of all, the third district including the coast line from New London, Conn., to Barnegat Inlet, New Jersey. This of course includes New York and adjacent commercial centres and the entrance to Long Island Sound with its long line of thriving cities and the ports of the places from which come our chief supplies of munitions of war. It includes the part of the United States which an enemy would most covet. The part which at once would furnish the richest plunder, and possession of which by a foe would most cripple this nation. To-day it is defended by stationary guns in land fortresses and in time of attack would be further guarded by a fringe of cruising naval vessels. Apparently up to the middle of 1917 the government thought no aerial watch was needed.

But if we were to follow the methods which all the belligerent nations of Europe are employing on their sea coasts we would establish in this district ten aeronautical stations. This would be no match for the British system which has one such station to every twenty miles of coast. Ours would be farther apart, but as the Sound could be guarded at its entrance the stations need only be maintained along the south shore of Long Island and down the Jersey coast. Each station would be provided with patrol, fighting, and observation aeroplanes. It would have the mechanical equipment of microphones, searchlights, and other devices for detecting the approach of an enemy now employed successfully abroad. Its patrolling aeroplanes would cruise constantly far out to sea, not less than eighty miles, keeping ever in touch with their station. As the horizon visible from a soaring aeroplane is not less than fifty miles distant from the observer, this would mean that no enemy fleet could approach within 130 miles of our coast without detection and report. The Montauk Point station would be charged with guarding the entrance to Long Island Sound and, the waters of Nantucket shoals and Block Island Sound where the German submarine *U-53* did its deadly work in 1916. The Sandy Hook station would of course be the most important of all, guarding New York sea-going commerce and protecting the ship channel by a constant patrol of aircraft over it.

The modern aeroplane has a speed of from eighty to one hundred and sixty miles an hour—the latter rate being attained only by the light scouts. Thus it is apparent that if an alarm were raised at any one of these stations between New London and Barnegat three hours at most would suffice to bring the fighting equipment of all the stations to the point threatened. There would be thus concentrated a fleet of several hundred swift scouts, heavy fighting machines, the torpedo planes of the type designed by Admiral Fiske, hydro-aeroplanes capable of carrying heavy guns and in brief every form of aerial fighter. Moreover, by use of the wireless, every ship of the Navy within a radius of several hundred miles would be notified of the menace. They could not reach the scene of action so swiftly as the flying men but the former would be able to hold the foe in action until the heavier ships should arrive.

The enormous advantage of such a system of guarding our coasts needs no further explanation. It is not even experimental, for France on her limited coast has 150 such stations. England, which started the war with 18, had 114 in 1917 and was still building. We at that time had none, although the extent of our sea coast and the great multiplicity of practicable harbours make us more vulnerable than any other nation.

CHAPTER 10
Some Features of Aerial Warfare

As devices to translate German hate for England into deeds of bloody malignancy and cowardly murder the German aircraft have ranked supreme. The ruthless submarine war has indeed done something toward working off this peculiar passion, but it lacked the spectacular qualities which German wrath demanded. As the war proceeded, and it became apparent that the participation of Great Britain—at first wholly unexpected by the Kaiser's advisers—was certain to defeat the German aims, the authorities carefully inculcated in the minds of the people the most malignant hatred for that power. As Lissauer's famous hymn of hate had it—

French and Russians it matters not,
A blow for a blow, and a shot for a shot.

We have one foe and one alone—England!

By way of at once gratifying this hatred and still further stimulating it the German military authorities began early in the war a series of air raids upon English towns. They were of more than doubtful military value. They damaged no military or naval works. They aroused the savage ire of the British people who saw their children slain in schools and their wounded in hospitals by bombs dropped from the sky and straightway rushed off to enlist against so callous and barbaric a foe. But the raids served their political purpose by making the German people believe that the British were suffering all the horrors of war on their own soil, while the iron line of trenches drawn across France by the German troops kept the invader and war's agonies far from the soil of the Fatherland.

The first German air raids were by Zeppelins on little English seaside towns—Scarborough, Hartlepool, and Harwich. Except in so

The U. S. Aviation School at Mineola

far as they inflicted mutilation and death upon many non-combatants, mostly women and children, and misery upon their relatives and friends they were without effect. But early in 1915 began a systematic series of raids upon London, which, by October of 1917, had totalled thirty-four, with a toll of 865 persons killed, and 2500 wounded. It seems fair to say that for these raids there was more plausible excuse than for those on the peaceful little seaside bathing resorts and fishing villages. London is full of military and naval centres, arsenals and navy yards, executive offices and centres of warlike activity. An incendiary bomb dropped into the Bank of England, or the Admiralty, might paralyze the finances of the Empire, or throw the naval organization into a state of anarchy. But as a matter of fact the German bombs did nothing of the sort. They fell in the congested districts of London, "the crowded warrens of the poor." They spread wounds and death among peaceable theatre audiences. One dropped on a 'bus loaded with passengers homeward bound, and obliterated it and them from the face of the earth. But no building of the least military importance sustained any injury. It is true, however, that the persistent raiding has compelled England to withhold from the fighting lines in France several thousand men and several hundred guns in order to be in readiness to meet air raids in which Germany has never employed more than fifty machines and at most two hundred men, including both aviators and mechanics.

It is entirely probable that the failure of the Germans to strike targets of military importance and the slaughter they wrought among peaceful civilians were due to no intent or purpose on their part. Hitting a chosen target from the air is no matter of certainty. The bomb intended for the railway station is quite as likely to hit the adjacent public school or hospital. If the world ever recurs to that moderate degree of sanity and civilization which shall permit wars, but strive to regulate them in the interest of humanity this untrustworthiness of the aircraft's aim will compel some form of international regulation, just as the vulnerability of the submarine will force the amendment of the doctrine of visitation and search. But neither problem can be logically and reasonably solved in the middle of a war. And so, while the German violation of existing international law had the uncomfortable result for Germany of bringing the United States into the war, the barbarous raids upon London caused the British at last to turn aside from their commendable abstention from air raids on unfortified and non-military towns and prepare for reprisals in kind.

From the beginning of the war the British had abstained from bombing peaceful and non-military towns. They had not indeed been weak in the employment of their air forces. General Smuts speaking in October, 1917, said that the British had, in the month previous, dropped 207 tons of bombs behind the lines of the enemy. But the targets were aerodromes, military camps, arsenals and munitions camps—not hospitals or kindergartens. The time had now come when this purely military campaign no longer satisfied an enraged British people who demanded the enforcement of the Mosaic law of an eye for an eye and a tooth for a tooth, against a people whom General Smuts described as "an enemy who apparently recognizes no laws, human or divine; who knows no pity or restraint, who sung *Te Deums* over the sinking of the *Lusitania*, and to whom the maiming and slaughter of women and children appear legitimate means of warfare."

And Premier Lloyd George, speaking to an audience of poor people in one of the congested districts which had suffered sorely from the aerial activities of the Hun, said:

"We will give it all back to them, and we will give it soon. We shall bomb Germany with compound interest."

But whether undertaken as part of a general programme of frightfulness or as reprisals for cruel and indefensible outrages air raids upon defenceless towns, killing peaceable citizens in their beds, and children in their kindergartens, are not incidents to add glory to aviation. The mind turns with relief from such examples of the cruel misuse of aircraft to the hosts of individual instances in which the airman and his machine remind one of the doughty Sir Knight and his charger in the most gallant days of chivalry. There were hosts of such incidents—men who fought gallantly and who always fought fair, men who hung about the outskirts of an aerial battle waiting for some individual champion of their own choosing to show himself and join in battle to death in the high ranges of the sky. Some of these have been mentioned in this book already. To discuss all who even as early as 1917 had made their names memorable would require a volume in itself. A few may well be mentioned below.

There, for example, was Captain Georges Guynemer, "King of the French Aces." An "ace" is an aviator who has brought down five enemy aircraft. Guynemer had fifty-three to his credit. Still a youth, only twenty-three years of age at the time of his death, and only flying for twenty-one months, he had lived out several life times in the mad excitement of combat in mid-air. Within three weeks after getting his aviator's license he had become an "Ace." Before his first year's service had expired he was decorated and promoted for gallantry in rushing to the aid of a comrade attacked by five enemy machines. He entered the combat at the height of ten thousand feet, and inside of two minutes had dropped two of the enemy. The others fled. He pursued hotly keeping up a steady fire with his machine gun. One Boche wavered and fell, but just then an enemy shell from an "Archie" far below exploded under Guynemer, tearing away one wing of his machine. Let him tell the rest of that story:

> I felt myself dropping. It was ten thousand feet to the earth, and, like a flash, I saw my funeral with my saddened comrades marching behind the gun carriage to the cemetery. But I pulled and pushed every lever I had, but nothing would check my terrific descent.
>
> Five thousand feet from the earth, the wrecked machine began to turn somersaults, but I was strapped into the seat. I do not know what it was, but something happened and I felt the speed descent lessen. But suddenly there was a tremendous crash and when I recovered my senses I had been taken from the wreckage and was all right.

Two records Guynemer made which have not yet been surpassed—the first, the one described above of dropping three Fokkers in two minutes and thirty seconds, and rounding off the adventure by himself dropping ten thousand feet. The second was in shooting down four enemy machines in one day. His methods were of the simplest. He was always alone in his machine, which was the lightest available. He would rather carry more gasoline and ammunition than take along a gunner. The machine gun was mounted on the plane above his head, pointing dead ahead, and aimed by aiming the whole aeroplane. Once started the gun continued firing automatically and Guynemer's task was to follow his enemy pitilessly keeping that lead-spitting muzzle steadily bearing upon him. In September, 1917, he went up to attack five enemy machines—no odds however appalling seemed to terrify him—but was caught in a fleet of nearly forty Boches and fell to earth in the enemy's country.

One of the last of the air duels to be fought under the practices which made early air service so vividly recall the age of chivalry, was that in which Captain Immelman, "The Falcon," of the German army, met Captain Ball of the British Royal Flying Corps. Immelman had a record of fifty-one British aeroplanes downed. Captain Ball was desirous of wiping out this record and the audacious German at the same time, and so flying over the German lines he dropped this letter:

> *Captain Immelman:*
> I challenge you to a man-to-man fight to take place this afternoon at two o'clock. I will meet you over the German lines. Have your anti-air craft guns withhold their fire, while we decide which is the better man. The British guns will be silent.
> *Ball*

Presently thereafter this answer was dropped from a German aeroplane:

> *Captain Ball:*
> Your challenge is accepted. The guns will not interfere. I will meet you promptly at two.
> *Immelman.*

The word spread far and wide along the trenches on both sides. Tacitly all firing stopped as though the bugles had sung truce. Men left cover and clambered up on the top to watch the duel. Punctually both flyers rose from their lines and made their way down No Man's Land. Let an eye witness tell the story:

> From our trenches there were wild cheers for Ball. The Germans yelled just as vigorously for Immelman.
>
> The cheers from the trenches continued; the Germans increased in volume; ours changed into cries of alarm.
>
> Ball, thousands of feet above us and only a speck in the sky, was doing the craziest things imaginable. He was below Immelman and was apparently making no effort to get above him, thus gaining the advantage of position. Rather he was swinging around, this way and that, attempting, it seemed, to postpone the inevitable.
>
> We saw the German's machine dip over preparatory to starting the nose dive.
>
> "He's gone now," sobbed a young soldier, at my side, for he knew Immelman's gun would start its raking fire once it was being driven straight down.
>
> Then in a fraction of a second the tables were turned. Before Immelman's plane could get into firing position, Ball drove

his machine into a loop, getting above his adversary and cutting loose with his gun and smashing Immelman by a hail of bullets as he swept by.

Immelman's aeroplane burst into flames and dropped. Ball, from above, followed for a few hundred feet and then straightened out and raced for home. He settled down, rose again, hurried back, and released a huge wreath of flowers, almost directly over the spot where Immelman's charred body was being lifted from a tangled mass of metal.

Four days later Ball too was killed.

But the Germans, too, had their champion airmen, mighty fliers, skilful at control and with the machine gun, in whose triumphs they took the same pride that our boys in France did in those of Chapman, Rockwell or Thaw, the British in Warneford, or the French in Guynemer. Chief of these was Captain Boelke, who came to his death in the latter part of 1917, after putting to his credit over sixty Allied planes brought down. A German account of one of his duels as watched from the trenches, will be of interest:

> For quite a long time an Englishman had been making circles before our eyes—calmly and deliberately.... My men on duty clenched their fists in impotent wrath. "The dog—!" Shooting would do no good.
>
> Then suddenly from the rear a harsh, deep singing and buzzing cuts the air. It sounds like a German flyer. But he is not yet visible. Only the buzz of an approaching motor is heard in the clouds in the direction of the Englishman. More than a hundred eyes scanned the horizon. There! Far away and high among the clouds is a small black humming bird—a German battle aeroplane. Its course is laid directly for the hostile biplane and it flies like an arrow shot with a clear eye and steady hand. My men crawl out of the shelters. I adjust my field glasses. A lump rises in our throats as if we are awaiting something new and wonderful.
>
> So far the other does not seem to have noticed or recognized the black flyer that already is poised as a hawk above him. All at once there is a mighty swoop through the air like the drop of a bird of prey, and in no time the black flyer is immediately over the Englishman and the air is filled with the furious crackling of a machine gun, followed by the rapid *ta-ta-ta* of two or three

more, all operated at the highest speed just as during a charge. The Englishman drops a little, makes a circle and tries to escape toward the rear. The other circles and attacks him in front, and again we hear the exciting *ta-ta-ta*! Now the Englishman tries to slip from under his opponent, but the German makes a circle and the effort fails. Then the enemy describes a great circle and attempts to rise above the German. The latter ascends in sharp half circles and again swoops down upon the biplane, driving it toward the German trenches.

Will the Englishman yield so soon? Scattered shouts of joy are already heard in our ranks. Suddenly he drops a hundred yards and more through the air and makes a skilful loop toward the rear. Our warrior of the air swoops after him, tackles him once more and again we hear the wild defiant rattle of the machine guns over our heads. Now they are quite close to our trenches. The French infantry and artillery begin firing in a last desperate hope. Neither of them is touched. Sticking close above and behind him the German drives the Englishman along some six hundred yards over our heads and then just above the housetops of St. A. Once more we hear a distant *ta-ta-ta* a little slower and more scattered and then as they drop both disappear from our view.

Scarcely five minutes pass before the telephone brings up this news: Lieutenant Boelke has just brought down his seventh flyer.

Methods of air-fighting were succinctly described in a hearing before the Senate Committee on Military Affairs, in June, 1917. The officers testifying were young Americans of the Lafayette Escadrille of the French army. To the civilian the testimony is interesting for the clear idea it gives of military aviation. The extracts following are from the official record:

> *Adjt. Prince*: Senator, there are about four kinds of machines used abroad on the western front to-day. The machines that Adjt. Rumsey and myself are looking after are called the battle machines. Then there are the photography machines, machines that go up to enable the taking of photographs of the German batteries, go back of the line and take views of the country behind their lines and find out what their next line of attack will be, or, if they retreat from the present line, then everything in that way. Probably we have, where we are, in my group alone, a

Miss Ruth Law at close of her Chicago to New York flight

hundred and fifty photographers who do nothing all day long except develop pictures, and you can get pictures of any part of the country that you want. When the Germans retreated from the old line where they used to be, by Peronne and Chaulnes, we had absolute pictures of all the Hindenburg line from where they are now right down to St. Quentin, down to the line the French are on. We had photographs of it all.

Senator Kirby: When they started on the retreat?

Adjt. Prince: Yes, sir. So we knew exactly where their stand would be made. Then, besides that, those photograph machines do a lot of scouting. They have a pilot and a photographer aboard. He has not only a camera, but quite often he has a Lewis gun with him in order to ward off any hostile airmen if they should get through the battle planes that are above him; in other words, should get through us in order to fight him. They do a great deal of the scouting, because they fly at a lower level. The battle planes go up to protect photography machines, or to go man-hunting, as it is called; in other words, to fight the Germans. We fly all day, like to-day, as high as we can go, or as high as the French go as a rule, about 5500 metres, about 17,000 to 18,000 feet.

Adjt. Rumsey: I think 5500 metres is about 19,000 feet. Some go up 6000 metres, which makes about 20,000 feet.

Adjt. Prince: We go up there, and we have a certain sector of the front to look after. If we are only man-hunting, we go backward and forward like a policeman to prevent the Germans from getting over our own lines. We usually fly by fours, if we can, and the four go out together, so as not to be alone. We are usually fighting inside of the German lines, because the morale of the French and English is better than that of the Germans to-day; and every fight I have had—I have never been lucky enough to have one inside of my own lines—they have all been inside of the German lines.

Senator Kirby: What is the equipment of a battle plane such as you use?

Adjt. Prince: I use the 180 horse-power machine. It is called a "S. P. A. D.," which has a Spanish motor. But a great many of the motors to-day are being built here in America.

Senator Kirby: How many men do you carry?

Adjt. Prince: We go up alone in these machines. We did have

A French aviator between flights

two guns. We had the Lewis gun on our upper wing and the Vickers down below, that shoots through the propeller as the propeller turns around. Then we gave up the Lewis above. It added more weight, and we did not need it so much. The trouble with the Lewis gun is that it has only ninety-seven cartridges, while the Vickers has five hundred, and you can do just as much damage with the Vickers as you could with them both.
Senator Sutherland: You drive and fight at the same time?
Adjt. Prince: Yes, sir.
Adjt. Rumsey: The machine gun is fixed.

Adjt. Prince: It is absolutely fixed on the machine, and if I should want to adjust it to shoot you, I would adjust my machine on you.

The witness then took up the nature and work of some of the heavier machines. He testified:

Adjt. Prince: Then comes the artillery regulating machine. That machine goes up, and it may be a Farman or a bi-motor, or some other kind of heavier machine, a machine that goes slowly. They go over a certain spot. They have a driver, who is a pilot, like ourselves; then they have an artillery officer on board, whose sole duty it is to send back word, mostly by Marconi, to his battery where the shots are landing. He will say: "Too far," "Too short," "Right," or "Left," and he stays there over this battery until the work done by the French guns has been absolutely controlled, and above him he has some of these battle planes keeping him from being attacked from above by German airmen. Of course, they may be shot at by anti-aircraft guns, which you can not help. That is artillery regulating.

The Chairman: Are you always attacked from above?

Adjt. Prince: By aeroplanes; yes, sir. It is always much safer to attack from above.

Then you have the bomb-dropping machines, which carry a lot of weight. They go out sometimes in the daytime, but mostly at night, and they have these new sights by which they can stay up quite high in the air and still know the spot they are going at. They know the wind speed, they know their height, and they can figure out by this new arrangement they have exactly when the time is to let go their bombs.

Senator Kirby: Something in the nature of a range-finder?

Adjt. Prince: A sort of range-finder.

Adjt. Rumsey: It is a sort of telescope that looks down between your legs, and you have to regulate yourself, observing your speed, and when you see the spot, you have to touch a button and off go these things.

Adjt. Rumsey: In a raid my brother went on there were sixty-eight machines that left; the French heavy machines, the English heavy machines, and then the English sort of half-fighting machine and half-bombing machine. They call it a Sopwith, and it is a very good machine. They went over there, and the first ones over were the Frenchmen, and they dropped bombs on these Mauser works, and the only thing that the English saw was a big cloud of smoke and dust, and they could not see the works so they just dropped into them. Out of that raid the fighting machines got eight Germans and dropped them, and the Germans got eight Frenchmen. So, out of sixty-eight they lost eight, but we also got eight Germans and dropped six tons of this stuff, which is twenty times as strong as the *melinite*. We do not know what the name of the powder is. The fighting machines on that trip only carried gasoline for two hours, and the other ones carried it for something like six hours, so we escorted them out for an hour, came back to our lines, filled up with gasoline, went out and met them and brought them back over the danger zone.

Adjt. Prince: Near the trenches is where the danger zone is, because there the German fighting machines are located.

Senator Kirby: How far was it from your battle front that you went?

Adjt. Rumsey: I think it was about 500 miles, 250 there and 250 back; it was between 200 and 250 miles there.

Senator Kirby: Beyond the battle front?

Adjt. Rumsey: Yes; or, to be more accurate, I think it was nearer 200 than 250.

The Chairman: What do you think of the function of the aeroplane as a determining factor?

Adjt. Prince: There is no doubt that if we could send over in huge waves a great number of these bomb-dropping machines, and simply lay the country waste—for instance, the big cities like Strassburg, Freiburg, and others—not only would the damage done be great, but I guess the popular opinion in Germany, everything being laid waste, would work very strongly in the minds of the public toward having peace. I do not think you could destroy an army, because you could not see them, but you could go to different stations; you could go to Strassburg, to Brussels, and places like that.

The Chairman: Then, sending them over in enormous numbers would also put out of business their aeroplanes, and they would be helpless, would they not?

Adjt. Prince: Absolutely. You not only have on the front a large number of bomb-dropping machines, but a large number of fighting machines. When the Somme battle was started in the morning the Germans knew, naturally, that the French and British were going to start the Somme drive, and they had up these Drachens, these observation balloons, and the first eighteen minutes that the battle started the French and the English, I think, got twenty-one "*saucisse*"; in other words, for the next five days there was not a single German who came anywhere near the lines, but the French and English could go ahead as they-felt like.

Admiral Peary: Have you any idea as to how many aeroplanes there are along that western front on the German side?

Adjt. Prince: There must be about 3000 on that line in actual commission.

Admiral Peary: That means, then, about 10,000 in all, at least?

Adjt. Prince: I should think so; I should say the French have about 2000 and the English possibly 1000, or we have about 2500.

Adjt. Rumsey: If they have 3000 we have 4000; that is, right on the line.

Adjt. Prince: We have about 1000 more than they have, and we are up all the time. The day before I left the front I was called to go out five times, and I went out five times, and spent two hours every time I went out.

It would be gratifying to author and to reader alike if it were possible to give some account of the progress in aerial equipment made by the United States, since its declaration of war. But at the present moment (February, 1918), the government is chary of furnishing information concerning the advance made in the creation of an aerial fleet. Perhaps precise information, if available, would be discouraging to the many who believe that the war will be won in the air. For it is known in a broad general way that the activities of the Administration have been centred upon the construction of training camps and aviation stations. Orders for the actual construction of aeroplanes have been limited, so that a chorus of criticism arose from manufacturers who declared that they might have to close their works for lack of employment. The apparent check was discouraging to American airmen, and to our Allies who had expected marvellous things from the United States in the way of swift and wholesale preparation for winning battles in the air. The response of the government to all criticism was that it was laying broad foundations in order that construction once begun would proceed with unabated activity, and that when aircraft began to be turned out by the thousands a week there would be aviators and trained mechanics a-plenty to handle them. In this situation the advocates of a special cabinet department of aeronautics found new reason to criticize the Administration and Congress for having ignored or antagonized their appeals. For responsibility for the delay and indifference—if indifference there was—rested equally upon the Secretary of the Navy and the Secretary of War. Each had his measure of control over the enormous sum voted in a lump for aviation, each had the further millions especially voted to his department to account for. But no single individual could be officially asked what had been done with the almost one billion dollars voted for aeronautics in 1917.

But if the authorities seemed to lag, the inventors were busy. Mention has already been made of the new "Liberty" motor, which report had it was the fruit of the imprisonment of two mechanical experts in a hotel room with orders that they should not be freed until they had produced a motor which met all criticisms upon those now in use. Their product is said to have met this test, and the happy result caused a general wish that the Secretaries of War and of the Navy might be similarly incarcerated and only liberated upon producing plans for the immediate creation of an aerial fleet suited to the nation's needs. If, however, the Liberty motor shall prove the complete success which at the moment the government believes it to be, it will be such a spur to the development of the aeroplane in peace and war, as could not otherwise be applied. For the motor is the true life of the aeroplane— its heart, lungs, and nerve centre. The few people who still doubt the wide adoption of aircraft for peaceful purposes after the war base their scepticism on the treachery of motors still in use. They repudiate all comparisons with automobiles. They say:

It is perfectly true that a man can run his car repeatedly from New York to Boston without motor trouble. But the trouble is inevitable sooner or later. When it comes to an automobile it is trifling. The driver gets out and makes his repairs by the roadside. But if it comes to the aviator it brings the possibility of death with it every time. If his motor stops he must descend. But to alight he must find a long level field, with at least two hundred yards in which to run off his momentum. If, when he discovers the failure of his motor, he is flying at the height of a mile he must find his landing place within a space of eight miles, for in gliding to earth the ratio of forward movement to height is as eight to one. But how often in rugged and densely populated New England, or Pennsylvania is there a vacant level field half a mile in length? The aviator who made a practice of daily flight between New York and Boston would inevitably meet death in the end.

The criticism is a shrewd and searching one. But it is based on the aeroplane and the motor of to-day without allowance for the development and improvement which are proceeding apace. It contemplates a craft which has but one motor, but the more modern machines have sufficient lifting power to carry two motors, and can be navigated successfully with one of these out of service. Experiments furthermore are being made with a device after the type of the helicopter which with the steady lightening of the aircraft motor, may be installed on aeroplanes with a special motor for its operation. This device, it is believed, will enable the aeroplane so equipped to stop dead in its course with both propellers out of action, to hover over a given spot or to rise or to descend gently in a perpendicular line without the necessity of soaring. It is obvious that if this device prove successful the chief force of the objections to aerial navigation outlined above will be nullified.

The menace of infrequent landing places will quickly remedy itself on busy lines of aerial traffic. The average railroad doing business in a densely populated section has stations once every eight or ten miles which with their sidings, buildings, water tanks, etc., cost far more than the field half a mile long with a few hangars that the fliers will need as a place of refuge. Indeed, although for its size and apparent simplicity of construction an aeroplane is phenomenally costly, in the grand total of cost an aerial line would cost a tithe of the ordinary railway. It has neither right of way, road bed, rails, nor telegraph system to maintain, and if the average flyer seems to cost amazingly it still foots up less than one fifth the cost of a modern locomotive though its period of service is much shorter.

Just at the present time aircraft costs are high, based on artificial conditions in the market. Their construction is a new industry; its processes not yet standardized; its materials still experimental in many ways and not yet systematically produced. A light sporting monoplane which superficially seems to have about $250 worth of materials in it—exclusive of the engine—will cost about $3000. A fighting biplane will touch $10,000. Yet the latter seems to the lay observer to contain no costly materials to justify so great a charge. The wings are a light wooden framework, usually of spruce, across which a fine grade of linen cloth is stretched. The materials are simple enough, but every bit of wood, every screw, every strand of wire is selected with the utmost care, and the workmanship of their assemblage is as painstaking as the setting of the most precious stones.

Remember the least negligence may cost a life! is a sign frequently seen hanging over the work benches in an aeroplane factory.

When stretched over the framework, the cloth of the wings is treated to a dressing down of a preparation of *collodion*, which in the jargon of the shop is called "dope." This substance has a peculiar effect upon the cloth, causing it to shrink, and thus making it more taut and rigid than it could be by the most careful stretching. Though the layman would not suspect it, this wash alone costs about $150 a machine. The seaplanes too—or hydro-aeroplanes as purists call them—present a curious illustration of unexpected and, it would seem, unexplainable expense. Where the flyer over land has two bicycle wheels on which to land, the flyer over the sea has two flat-bottomed boats or pontoons. These cost from $1000 to $1200 and look as though they should cost not over $100. But the necessity of combining maximum strength with minimum weight sends the price soaring as the machine itself soars. Moreover there is not yet the demand for either air-or seaplanes that would result in the division of labour, standardization of parts, and other manufacturing economies which reduce the cost of products.

To the high cost of aircraft their comparative fragility is added as a reason for their unfitness for commercial uses. The engines cost from $2000 to $5000 each, are very delicate and usually must be taken out of the plane and overhauled after about 100 hours of active service.

A German Gotha—their favourite type

The strain on them is prodigious for it is estimated that the number of revolutions of an aeroplane's engine during an hour's flight is equal to the number of revolutions of an automobile's wheels during active service of a whole month.

It is believed that the superior lightness and durability of the Liberty motor will obviate some of these objections to the commercial availability of aircraft in times of peace. And it is certain that with the cessation of the war, the retirement of the governments of the world from the purchasing field and the reduction of the demand for aircraft to such as are needed for pleasure and industrial uses the prices which we have cited will be cut in half. In such event what will be the future of aircraft; what their part in the social and industrial organization of the world?

Ten or a dozen years ago Rudyard Kipling entertained the English reading public of the world with a vivacious sketch of aerial navigation in the year 2000 A.D. He used the license of a poet in avoiding too precise descriptions of what is to come—dealing rather with broad and picturesque generalizations. Now the year 2000 is still far enough away for pretty much anything to be invented, and to become commonplace before that era arrives. Airships of the sort Mr. Kipling pictured may by that period have come and gone—have been relegated to the museums along with the stage-coaches of yesterday and the locomotives of to-day. For that matter before that millennial period shall arrive men may have learned to dispense with material transportation altogether, and be able to project their consciousness or even their astral bodies to any desired point on psychic waves. If a poet is going to prophecy he might as well be audacious and even revolutionary in his predictions.

Mr. Kipling tried so hard to be reasonable that he made himself recognizably wrong so far as the present tendency of aircraft development would indicate. *With the Night Mail,* is the story of a trip by night across the Atlantic from England to America. It is made in a monster dirigible—though the present tendency is to reject the dirigible for the swifter, less costly, and more airworthy (leave "seaworthy" to the plodding ships on old ocean's breast) aeroplanes. If, however, we condone this glaring improbability we find Mr. Kipling's tale full of action and imaginary incident that give it an air of truth. His ship is not docked on the ground at the tempest's mercy, but is moored high in air to the top of a tall tower up which passengers and freight are conveyed in elevators. His lighthouses send their beams straight up into the sky instead of projecting them horizontally as do those which now guard our coasts. Just why lighthouses are needed, however, he does not explain. There are no reefs on which a packet of the air may run, no lee shores which they must avoid. On overland voyages guiding lights by night may be useful, as great white direction strips laid out on the ground are even now suggested as guides for daylight flying. But the main reliance of the airman must be his compass. Crossing the broad oceans no lighted path is possible, and even in a voyage from New York to Chicago, or from London to Rome good airmanship will dictate flight at a height that will make reliance upon natural objects as a guide perilous. The airman has the advantage over the sailor in that he may lay his course on leaving his port, or flying field, and pursue it straight as an arrow to his destination. No rocks or other obstacles bar his path, no tortuous channels must be navigated. All that can divert him from his chosen course is a steady wind on the beam, and that is instantly detected by his instruments and allowance made for it. On the other hand the sailor has a certain advantage over the airman in that his more leisurely progress allows time for the rectification of errors in course arising from contrary currents or winds. An error of a point, or even two, amounts to but little in a day's steaming of perhaps four hundred miles. It can readily be remedied, unless the ship is too near shore. But when the whole three thousand miles of Atlantic are covered in twenty hours in the air, the course must be right from the start and exactly adhered to, else the passenger for New York may be set down in Florida.

It is not improbable that even before the war is over the crossing of the Atlantic by plane will be accomplished. Certainly it will be one of the first tasks undertaken by airmen on the return of peace. But it is probable that the adaptation of aircraft to commercial uses will be begun with undertakings of smaller proportions. Already the United States maintains an aerial mail route in Alaska, while Italy has military mail routes served by aeroplanes in the Alps. These have been undertaken because of the physical obstacles to travel on the surface, presented in those rugged neighbourhoods. But in the more densely populated regions of the United States considerations of financial profit will almost certainly result in the early establishment of mail and passenger air service. Air service will cut down the time between any two given points at least one half, and ultimately two thirds. Letters could be sent from New York to Boston, or even to Buffalo, and an answer received the same day. The carrying plane could take on each trip five tons of mail. Philadelphia would be brought within forty-five

A French monoplane

minutes of New York; Washington within two hours instead of the present five. Is there any doubt of the creation of an aerial passenger service under such conditions? Already a Caproni triplane will carry thirty-five passengers beside guns—say, fifty passengers if all other load be excluded, and has flown with a lighter load from Newport News to New York. It is easily imaginable that by 1920 the aeroplane capable of carrying eighty persons—or the normal number now accommodated on an inter-urban trolley car—will be an accomplished fact.

The lines that will thus spring up will need no rails, no right of way, no expensive power plant. Their physical property will be confined to the aeroplanes themselves and to the fields from which the craft rise and on which they alight, with the necessary hangars. These indeed will involve heavy expenditure. For a busy line, with frequent sailings, of high speed machines a field will need to be in the neighbourhood of a mile square. A plane swooping down for its landing is not to be held up at the switch like a train while room is made for it. It is an imperative guest, and cannot be gainsaid. Accordingly the fields must be large enough to accommodate scores of planes at once and give each new arrival a long straight course on which to run off its momentum. It is obvious therefore that the union stations for aircraft routes cannot be in the hearts of our cities as are the railroad stations of to-day, but must be fairly well out in the suburbs.

A form of machine which the professional airmen say has yet to be developed is the small monoplane, carrying two passengers at most, and of low speed—not more than twenty miles an hour at most. In this age of speed mania the idea of deliberately planning a conveyance or vehicle that shall not exceed a low limit seems out of accord with public desire. But the low speed aeroplane has the advantage of needing no extended field in which to alight. It reaches the ground with but little momentum to be taken up and can be brought up standing on the roof of a house or the deck of a ship. Small machines of this sort are likely to serve as the runabouts of the air, to succeed the trim little automobile roadsters as pleasure craft.

The beginning of the fourth year of the war brought a notable change in aerial tactics. For three years everything had been sacrificed to speed. Such aerial duels as have been described were encouraged by the fact that aircraft were reduced to the proportions needful for carrying one man and a machine gun. The gallant flyers went up in the air and killed each other. That was about all there was to it. While as scouts, range finders, guides for the artillery, they exerted some in-

A German scout brought to earth in France

fluence on the course of the war, as a fighting arm in its earlier years, they were without efficiency. The bombing forays were harassing but little more, because the craft engaged were of too small capacity to carry enough bombs to work really serious damage, while the ever increasing range of the "Archies" compels the airmen to deliver their fire from so great a height as to make accurate aim impossible.

But Kiel, Wilhelmshaven and Zeebrugge are likely to change all this. The constant contemplation of those nests for the sanctuary of pestiferous submarines, effectively guarded against attack by either land or water, has stirred up the determination of the Allies to seek their destruction from above. Heavy bombing planes are being built in all the Allied workshops for this purpose, and furthermore to give effect to the British determination to take vengeance upon Germany, for her raids upon London. It is reported that the United States, by agreement with its Allies, is to specialize in building the light, swift scout planes, but in other shops the heavy triplane, the dreadnought of the air is expected to be the feature of 1918. With it will come an entirely novel strategic use of aircraft in war, and with it too, which is perhaps the more permanently important, will come the development of aircraft of the sort that will be readily adaptable to the purposes of peace when the war shall end.

The Submarine Boat

CHAPTER 11

Beginnings of Submarine Invention

In September, 1914 the British Fleet in the North Sea had settled down to the monotonous task of holding the coasts of Germany and the channels leading to them in a state of blockade. The work was dismal enough. The ships tossing from day to day on the always unquiet waters of the North Sea were crowded with Jackies all of whom prayed each day that the German would come from hiding and give battle. Not far from the Hook of Holland engaged in this monotonous work were three cruisers of about 12,000 tons, each carrying 755 men and officers. They were the *Cressy*, *Aboukir*, and *Hogue*—not vessels of the first rank but still important factors in the British blockade. They were well within the torpedo belt and it may be believed that unceasing vigilance was observed on every ship. Nevertheless without warning the other two suddenly saw the *Aboukir* overwhelmed by a flash of fire, a pillar of smoke and a great geyser of water that rose from the sea and fell heavily upon her deck. Instantly followed a thundering explosion as the magazines of the doomed ship went off. Within a very few minutes, too little time to use their guns against the enemy had they been able to see him, or to lower their boats, the *Aboukir* sank leaving the crew floundering in the water.

In the distance lay the German submarine *U-9*—one of the earliest of her class in service. From her conning tower Captain Weddigen had viewed the tragedy. Now seeing the two sister ships speeding to the rescue he quickly submerged. It may be noted that as a result of what followed, orders were given by the British Admiralty that in the event of the destruction of a ship by a submarine others in the same squadron should not come to the rescue of the victim, but scatter as widely as possible to avoid a like fate. In this instance the *Hogue* and the *Cressy* hurried to the spot whence the *Aboukir* had vanished and began lowering their boats. Hardly had they begun the work of mercy when a torpedo from the now unseen foe struck the *Hogue* and in twenty minutes she too had vanished. While she was sinking the *Cressy*, with all guns ready for action and her gunners scanning the sea in every direction for this deadly enemy, suddenly felt the shock of a torpedo and, her magazines having been set off, followed her sister ships to the ocean's bed.

In little more than half an hour thirty-six thousand tons of up-to-date British fighting machinery, and more than 1200 gallant blue jackets had been sent to the depths of the North Sea by a little boat of 450 tons carrying a crew of twenty-six men.

The world stood aghast. With the feeling of horror at the swift death of so many caused by so few, there was mingled a feeling of amazement at the scientific perfection of the submarine, its power, and its deadly work. Men said it was the end of dreadnoughts, battleships, and cruisers, but the history of the war has shown singularly few of these destroyed by submarines since the first novelty of the attack wore off. The world at the moment seemed to think that the submarine was an entirely new idea and invention. But like almost everything else it was merely the ultimate reduction to practical use of an idea that had been germinating in the mind of man from the earliest days of history.

We need not trouble ourselves with the speculations of Alexander the Great, Aristotle, and Pliny concerning "underwater" activities. Their active minds gave consideration to the problem, but mainly as to the employment of divers. Not until the first part of the sixteenth century do we find any very specific reference to actual underwater boats. That appears in a book of travels by Olaus Magnus, Archbishop of Upsala in Sweden. Notwithstanding the gentleman's reverend quality, one must question somewhat the veracity of the chapter which he heads:

Of the Leather Ships Made of Hides Used by the Pyrats of Greenland.

He professed to have seen two of these "ships," more probably boats, hanging in a cathedral church in Greenland. With these singular vessels, according to his veracious reports the people of that country could navigate under water and attack stranger ships from beneath.

> For the Inhabitants of that Countrey are wont to get small profits by the spoils of others, by these and the like treacherous Arts, who by their thieving wit, and by boring a hole privately in the sides of the ships beneath (as I said) have let in the water and presently caused them to sink.

Leaving the tale of the Archbishop where we think it must belong in the realm of fiction, we may note that it was not until the beginning of the seventeenth century that the first submarine boat was actually built and navigated. A Hollander, Cornelius Drebel, or Van Drebel, born in 1572, in the town of Alkmaar, had come to London during the reign of James I., who became his patron and friend. Drebel seems to have been a serious student of science and in many ways far ahead of his times. Moreover, he had the talent of getting next to royalty. In 1620 he first conceived the idea of building a submarine. Fairly detailed descriptions of his boats—he built three from 1620-1624—and of their actual use, have been handed down to us by men whose accuracy and truthfulness cannot be doubted. The Honourable Robert Boyle, a scientist of unquestioned seriousness, tells in his *New Experiments, Physico-Mechanical touching the Spring of the Air and its Effects* about Drebel's work in the quaint language of his time:

> But yet on occasion of this opinion of Paracelsus, perhaps it will not be impertinent if, before I proceed, I acquaint your Lordship with a conceit of that deservedly famous mechanician and Chymist, Cornelius Drebel, who, among other strange things that he perform'd, is affirm'd, by more than a few credible persons, to have contrived for the late learned King James, a vessel to go under water; of which, trial was made in the Thames, with admired success, the vessel carrying twelve rowers, besides passengers; one which is yet alive, and related it to an excellent Mathematician that informed me of it. Now that for which I mention this story is, that having had the curiosity and opportunity to make particular inquiries among the relations of Drebel, and especially of an ingenious physician that married his daughter, concerning the grounds upon which he conceived it feasible to make men unaccustomed to continue so long under water without suffocation, or (as the lately mentioned person that went in the vessel affirms) without inconvenience; I was answered, that Drebel conceived, that it is not the whole body of the air, but a certain quintessence (as Chymists speak) or spirituous part of it, that makes it fit for respiration; which being spent, the remaining grosser body, or carcase, if I may so call it, of the air, is unable to cherish the vital flame residing in the heart; so that, for aught I could gather, besides the mechanical contrivances of his vessel, he had a chymi-

cal liquor, which he accounted the chief secret of his submarine navigation. For when, from time to time, he conceived that the finer and purer part of the air was consumed, or over-clogged by the respiration and steam of those that went in his ship, he would by unstopping a vessel full of this liquor, speedily restore to the troubled air such a proportion of vital parts, as would make it again, for a good while, fit for respiration whether by dissipating, or precipitating the grosser exhalations, or by some other intelligible way, I must not now stay to examine, contenting myself to add, that having had the opportunity to do some service to those of his relations that were most intimate with him, and having made it my business to learn what this strange liquor might be, they constantly affirmed that Drebel would never disclose the liquor unto any, nor so much as tell the nature whereof he had made it, to above one person, who himself assured me what it was.

This most curious narrative suggests that in some way Drebel, who died in London in 1634, had discovered the art of compressing oxygen and conceived the idea of making it serviceable for freshening the air in a boat, or other place, contaminated by the respiration of a number of men for a long time. Indeed the reference made to the substance by which Drebel purified the atmosphere in his submarine as "a liquor" suggests that he may possibly have hit upon the secret of liquid air which late in the nineteenth century caused such a stir in the United States. Of his possession of some such secret there can be no doubt whatsoever, for Samuel Pepys refers in his famous diary to a lawsuit, brought in the King's Courts by the heirs of Drebel, to secure the secret for their own use. What was the outcome of the suit or the subsequent history of Drebel's invention history does not record.

Throughout the next 150 years a large number of inventors and near-inventors occupied themselves with the problem of the submarine. Some of these men went no further than to draw plans and to write out descriptions of what appeared to them to be feasible submarine boats. Others took one step further, by taking out patents, but only very few of the submarine engineers of this period had either the means or the courage to test their inventions in the only practicable way, by building an experimental boat and using it.

In spite of this apparent lack of faith on the part of the men who worked on the submarine problem, it would not be fair to condemn

A GAS ATTACK PHOTOGRAPHED FROM AN AEROPLANE

them as fakirs. Experimental workers, in those times, had to face many difficulties which were removed in later times. The study of science and the examination of the forces of nature were not only not as popular as they became later, but frequently were looked upon as blasphemous, savouring of sorcery, or as a sign of an unbalanced mind.

England and France supplied most of the men who occupied themselves with the submarine problem between 1610 and 1760. Of the Englishmen, the following left records of one kind or another concerning their labours in this direction. Richard Norwood, in 1632, was granted a patent for a contrivance which was apparently little more than a diving apparatus. In 1648, Bishop Wilkins published a book, *Mathematical Magick*, which was full of rather grotesque projects and which contained one chapter on the possibility "of framing an ark for submarine navigation." In 1691, patents were granted on engines connected with submarine navigation to John Holland—curious forerunner of a name destined to be famous two hundred years later—and on a submarine boat to Sir Stephen Evance.

In Prance, two priests, Fathers Mersenne and Fournier, published in 1634 a small book called *Questions Théologiques, Physiques, Morales et Mathématiques*, which contained a detailed description of a submarine boat. They suggested that the hull of submarines ought to be of metal and not of wood, and that their shape ought to be as nearly fishlike as possible. Nearly three hundred years have hardly altered these opinions. Ancient French records also tell us that six years later, in 1640, the King of France had granted a patent to Jean Barrié, permitting him during the next twelve years to fish at the bottom of the sea with his boat. Unluckily Barrié's fish stories have expired with his permit. In 1654, a French engineer, De Son, is said to have built at Rotterdam a submarine boat. Little is known concerning this vessel except that it was reported to have been seventy-two feet long, twelve feet high, and eight feet broad, and to have been propelled by a paddlewheel instead of oars.

Borelli, about whom very little seems to be known, is credited with having invented in 1680 a submarine boat, whose descent and ascent were regulated by a series of leather bottles placed in the hull of the boat with their mouths open to the surrounding water. The English magazine, *Graphic*, published a picture which is considered the oldest known illustration of any submarine boat. This picture matches in all details the description of Borelli's boat, but it is credited to a man called Symons.

Twenty-seven years later, in 1774, another Englishman, J. Day, built a small submarine boat, and after fairly extensive experiments, descended in his boat in Plymouth harbour. This descent is of special interest because we have a more detailed record of it than of any previous submarine exploit, and because Day is the first submarine inventor who lost his life in the attempt to prove the feasibility of his invention. The *Annual Register* of 1774 gives a narration in detail of Day's experiments and death and inasmuch as this is the first ungarbled report of a submarine descent, it may be quoted at length.

AUTHENTIC ACCOUNT OF A LATE UNFORTUNATE TRANSACTION, WITH RESPECT TO A DIVING MACHINE AT PLYMOUTH

Mr. Day (the sole projector of the scheme, and, as matters have turned out, the unhappy sacrifice to his own ingenuity) employed his thoughts for some years past in planning a method of sinking a vessel under water, with a man in it, who should live therein for a certain time, and then by his own means only, bring himself up to the surface. After much study he conceived that his plan could be reduced into practice. He communicated his idea in the part of the country where he lived, and had the most sanguine hopes of success. He went so far as to try his project in the Broads near Yarmouth. He fitted a Norwich market-boat for his purpose, sunk himself thirty feet under water, where he continued during the space of twenty-four hours, and executed his design to his own entire satisfaction. Elated with this success, he then wanted to avail himself of his invention. He conversed with his friends, convinced them that he had brought his undertaking to a certainty; but how to reap the advantage of it was the difficulty that remained. The person in whom he confided suggested to him, that, if he acquainted the sporting Gentlemen with the discovery, and the certainty of the performance, considerable betts would take place, as soon as the project would be mentioned in company. The Sporting Kalendar was immediately looked into, and the name of Blake soon occurred; that gentleman was fixed upon as the person to whom Mr. Day ought to address himself. Accordingly, Mr. Blake, in the month of November last, received the following letter:

Sir,

I found out an affair by which many thousands may be won; it is of a paradoxical nature, but can be performed with ease; therefore, sir, if you chuse to be informed of it, and give me one hundred pounds of every thousand you shall win by it, I will very readily wait upon you and inform you of it. I am myself but a poor mechanic and not able to make anything by it without your assistance.
Yours, etc.

J. Day

Mr. Blake had no conception of Mr. Day's design, nor was he sure that the letter was serious. To clear the matter up, he returned for answer, that, if Mr. Day would come to town, and explain himself, Mr. Blake would consider of the proposal. If he approved of it, Mr. Day should have the recompense he desired; if, on the other hand, the plan should be rejected, Mr. Blake would make him a present to defray the expenses of his journey. In a short time after Mr. Day came to town; Mr. Blake saw him and desired to know what secret he was possessed of. The man replied, "that he could sink a ship 100 feet deep in the sea with himself in it, and remain therein for the space of 24 hours, without communication with anything above; and at the expiration of the time, rise up again in the vessel." The proposal, in all its parts, was new to Mr. Blake. He took down the particulars, and, after considering the matter, desired some kind of proof of the practicability. The man added that if Mr. Blake would furnish him with the materials necessary, he would give him an ocular demonstration. A model of the vessel, with which he was to perform the experiment, was then required, and in three or four weeks accomplished, so as to give a perfect idea of the principle upon which the scheme was to be executed, and, in time, a very plausible promise of success, not to Mr. Blake only, but many other gentlemen who were consulted upon the occasion. The consequence was, that Mr. Blake, agreeably to the man's desire, advanced money for the construction of a vessel fit for that purpose. Mr. Day, thus assisted, went to Plymouth with his model, and set a man in that place to work upon it. The pressure of the water at 100 feet deep was a circumstance of which Mr. Blake was advised, and touching that article he gave

A FRENCH NIEUPORT DROPPING A BOMB

the strongest precautions to Mr. Day, telling him, at any expense, to fortify the chamber in which he was to subsist, against the weight of such a body of water. Mr. Day set off in great spirits for Plymouth, and seemed so confident, that Mr. Blake made a bett that the project would succeed, reducing, however, the depth of water from 100 yards to 100 feet, and the time from 24 to 12 hours. By the terms of the wager, the experiment was to be made within three months from the date; but so much time was necessary for due preparation, that on the appointed day things were not in readiness and Mr. Blake lost the bett.

In some short time afterwards the vessel was finished, and Mr. Day still continued eager for the carrying of his plan into execution; he was uneasy at the idea of dropping the scheme and wished for an opportunity to convince Mr. Blake that he could perform what he had undertaken. He wrote from Plymouth that everything was in readiness and should be executed the moment Mr. Blake arrived. Induced by this promise, Mr. Blake set out for Plymouth; upon his arrival a trial was made in Cat-water, where Mr. Day lay, during the flow of tide, six hours, and six more during the tide of ebb; confined all the time in the room appropriated for his use. A day for the final determination was fixed; the vessel was towed to the place agreed upon; Mr. Day provided himself with whatever he thought necessary; he went into the vessel, let the water into her and with great composure retired to the room constructed for him, and shut up the valve. The ship went gradually down in 22 fathoms of water at 2 o'clock on Tuesday, June 28, in the afternoon, being to return at 2 the next morning. He had three buoys or messengers, which he could send to the surface at option, to announce his situation below; but, none appearing, Mr. Blake, who was near at hand in a barge, began to entertain some suspicion. He kept a strict lookout, and at the time appointed, neither the buoys nor the vessel coming up, he applied to the *Orpheus* frigate, which lay just off the barge, for assistance. The captain with the most ready benevolence supplied them with everything in his power to seek for the ship. Mr. Blake, in this alarming situation was not content with the help of the *Orpheus* only; he made immediate application to Lord Sandwich (who happened to be at Plymouth) for further relief. His Lordship with great humanity ordered a number of hands from the dock-yard, who went with the utmost alacrity and tried every effort to regain the ship, but unhappily without effect.

Thus ended this unfortunate affair. Mr. Blake had not experience enough to judge of all possible contingencies, and he had now only to lament the credulity with which he listened to a projector, fond of his own scheme but certainly not possessed of skill enough to guard against the variety of accidents to which he was liable. The poor man has unfortunately shortened his days; he was not however tempted or influenced by anybody; he confided in his own judgment, and put his life to the hazard upon his own mistaken notions.

Many and various have been the opinions on this strange, useless, and fatal experiment, though the more reasonable part of mankind seemed to give it up as wholly impracticable. It is well-known, that pent-up air, when overcharged with the vapours emitted out of animal bodies, becomes unfit for respiration; for which reason, those confined in the diving-bell, after continuing some time under water are obliged to come up, and take in fresh air, or by some such means recruit it. That any man should be able after having sunk a vessel to so great a depth, to make that vessel at pressure, so much more specifically lighter than water, as thereby to enable it to force its way to the surface, through the depressure of so great a weight, is a matter not hastily to be credited. Even cork, when sunk to a certain depth will, by the great weight of the fluid upon it, be prevented from rising.

The English of the *Annual Register* leaves much to be desired in clarity. It makes reasonably clear, however, that the unfortunate Mr. Day's knowledge of submarine conditions was, by no means, equal to Mr. Blake's sporting spirit. Even to-day one hundred feet is an unusual depth of submersion for the largest submarines.

The credit for using a submarine boat for the first time in actual warfare belongs to a Yankee, David Bushnell. He was born in Saybrook, Connecticut, and graduated from Yale with the class of 1775. While still in college he was interested in science and as far as his means and opportunities allowed, he devoted a great deal of his time and energy to experimental work. The problem which attracted his special attention was how to explode powder under water, and before very long he succeeded in solving this to his own satisfaction as well as to that of a number of prominent people amongst whom were the Governor of Connecticut and his Council. Bushnell's experiments, of course, fell in the period during which the Revolutionary War was fought, and when he had completed his invention, there naturally presented itself to him a further problem. How could his device be used for the benefit of his country and against the British ships which were then threatening New York City? As a means to this end, Bushnell planned and built a submarine boat which on account of its shape is usually called the *Turtle*.

General Washington thought very highly of Bushnell, whom he called in a letter to Thomas Jefferson "a man of great mechanical powers, fertile in inventions and master of execution." In regard to Bushnell's submarine boat the same letter, written after its failure, says: "I thought and still think that it was an effort of genius, but that too many things were necessary to be combined to expect much against an enemy who are always on guard."

During the whole period of the building of the *Turtle* Bushnell was in ill health. Otherwise he would have navigated it on its trial trip himself for he was a man of undoubted courage and wrapped up alike in the merits of his invention and in the possibility of utilizing it to free New York from the constant ignominy of the presence of British ships in its harbour. But his health made this out of the question. Accordingly he taught his brother the method of navigating the craft, but at the moment for action the brother too fell ill. It became necessary to hire an operator. This was by no means easy as volunteers to go below the water in a submarine boat of a type hitherto undreamed of, and to attach an explosive to the hull of a British man-of-war, the sentries upon which were presumably especially vigilant, being in a hostile harbour, was an adventure likely to attract only the most daring and reckless spirits. In a letter to Thomas Jefferson, other portions of which we shall have occasion to quote later, Bushnell refers to this difficulty in finding a suitable operator and tells briefly and with evident chagrin the story of the failure of the attempts made to utilize successfully his submarine:

> After various attempts to find an operator to my wish, I sent one who appeared more expert than the rest from New York to a 50-gun ship lying not far from Governor's Island. He went under the ship and attempted to fix the wooden screw

A BOMB-DROPPING TAUBE

into her bottom, but struck, as he supposes, a bar of iron which passes from the rudder hinge, and is spiked under the ship's quarter. Had he moved a few inches, which he might have done without rowing, I have no doubt but he would have found wood where he might have fixed the screw, or if the ship were sheathed with copper he might easily have pierced it; but, not being well skilled in the management of the vessel, in attempting to move to another place he lost the ship. After seeking her in vain for some time, he rowed some distance and rose to the surface of the water, but found daylight had advanced so far that he durst not renew the attempt. He says that he could easily have fastened the magazine under the stem of the ship above water, as he rowed up to the stern and touched it before he descended. Had he fastened it there the explosion of 150 lbs. of powder (the quantity contained in the magazine) must have been fatal to the ship. In his return from the ship to New York he passed near Governor's Island, and thought he was discovered by the enemy on the island. Being in haste to avoid the danger he feared, he cast off the magazine, as he imagined it retarded him in the swell, which was very considerable. After the magazine had been cast off one hour, the time the internal apparatus was set to run, it blew up with great violence.

Afterwards there were two attempts made in Hudson's River, above the city, but they effected nothing. One of them was by the aforementioned person. In going towards the ship he lost sight of her, and went a great distance beyond her. When he at length found her the tide ran so strong that, as he descended under water for the ship's bottom, it swept him away. Soon after this the enemy went up the river and pursued the boat which had the submarine vessel on board and sunk it with their shot. Though I afterwards recovered the vessel, I found it impossible at that time to prosecute the design any farther.

The operator to whom Bushnell had entrusted his submarine boat was a typical Yankee, Ezra Lee of Lyme, Connecticut. His story of the adventure differs but little from that of Bushnell, but it is told with a calm indifference to danger and a seeming lack of any notion of the extraordinary in what he had done that gives an idea of the man.

> When I rode under the stern of the ship (the *Eagle*) I could see the men on deck and hear them talk, I then shut down all the doors, sunk down, and came up under the bottom of the ship.

This means that he hermetically sealed himself inside of a craft, shaped like two upper turtle shells joined together—hence the name of the *Turtle*. He had entered through the orifice at the top, whence the head of the turtle usually protrudes. This before sinking he had covered and made water-tight by screwing down upon it a brass crown or top like that to a flask. Within he had enough air to support him thirty minutes. The vessel stood upright, not flat as a turtle carries himself. It was maintained in this position by lead ballast. Within the operator occupied an upright position, half sitting, half standing. To sink water was admitted, which gathered in the lower part of the boat, while to rise again this was expelled by a force pump. There were ventilators and portholes for the admission of light and air when operating on the surface, but once the cap was screwed down the operator was in darkness.

In this craft, which suggests more than anything else a curiously shaped submarine coffin, Lee drifted along by the side of the ship, navigating with difficulty with his single oar and seeking vainly to find some spot to which he might affix his magazine. A fact which might have disquieted a more nervous man was that the clockwork of this machine was running and had been set to go off in an hour from the time the voyage was undertaken. As to almost anyone in that position minutes would seem hours, the calmness of sailor Lee's nerves seems to be something beyond the ordinary.

When he finally abandoned the attempt on the *Eagle* he started up the bay. Off Governor's Island he narrowly escaped capture.

> When I was abreast of the Fort on the Island three hundred or four hundred men got upon the parapet to observe me; at length a number came down to the shore, shoved off a twelve oar'd barge with five or six sitters and pulled for me. I eyed them, and when they had got within fifty or sixty yards of me I let loose the magazine in hopes that if they should take me they would likewise pick up the magazine and then we should all be blown up together. But as kind providence would have it they took fright and returned to the Island to my infinite joy.... The magazine after getting a little past the Island went off with a tremendous explosion, throwing up large bodies of water to an immense height.

During the last quarter of the eighteenth and during the first half of the nineteenth century France was the chief centre for the activities of submarine inventors. However, very few of the many plans put forward in this period were executed. The few exceptions resulted in little else than trial boats which usually did not live up to the expectations of their inventors or their financial backers and were, therefore, discarded in quick order. In spite of this lack of actual results this particular period was of considerable importance to the later development of the submarine. Almost every one of the many boats then projected or built contained some innovation and in this way some of the many obstacles were gradually overcome. Strictly speaking the net result of the experimental work done during these seventy-five years by a score or more of men, most of whom were French, though a few were English, was the creation of a more sane and sound basis on which, before long, other men began to build with greater success.

The one notable accomplishment of interest, especially to Americans, was the submarine built in 1800-01 by Robert Fulton. Fulton, of course, is far better known by his work in connection with the discovery and development of steam navigation. Born in Pennsylvania in 1765, he early showed marked mechanical genius. In 1787 he went to England with the purpose of studying art under the famous painter West, but soon began to devote most of his time and energy to mechanical problems. Not finding in England as much encouragement as he had hoped, he went, in 1797, to Paris and, for the next seven years, lived there in the house of the American Minister, Joel Barlow.

As soon as he had settled down in France, he offered his plans of a submarine boat which he called the *Nautilus* to the French Government. Though a special commission reported favourably on this boat, the opposition of the French Minister of the Marine was too strong to be overcome, even after another commission had approved a model built by Fulton. In 1800, however, he was successful in gaining the moral and financial support of Napoleon Bonaparte, then First Consul of the French Republic.

Fulton immediately proceeded to build the *Nautilus* and completed the boat in May, 1801. It was cigar-shaped, about seven feet in diameter and over twenty-one feet in length. The hull was of copper strengthened by iron ribs. The most noticeable features were a collapsible mast and sail and a small conning tower at the forward end. The boat was propelled by a wheel affixed to the centre of the stern

A CAPTURED GERMAN FOKKER EXHIBITED AT THE INVALIDES

and worked by a hand-winch. A rudder was used for steering, and increased stability was gained by a keel which ran the whole length of the hull.

Soon after completion the boat was taken out for a number of trial trips all of which were carried out with signal success and finally culminated, on June 26, 1801, in the successful blowing up of an old ship furnished by the French Government. Although the *Nautilus* created a great sensation, popular as well as official interest began soon to flag. Fulton received no further encouragement and finally gave up his submarine experiments.

In 1806 he returned to America. By 1814 he had built another submarine boat which he called the *Mute*. It was, comparatively speaking, of immense size, being over eighty feet long, twenty-one feet wide, and fourteen feet deep and accommodating a hundred men. It was iron-plated on top and derived its peculiar name from the fact that it was propelled by a noiseless engine. Before its trials could be completed, Fulton died on February 24, 1815, and no one seemed to have sufficient interest or faith in his new boat to continue his work.

In the middle of the nineteenth century for the first time a German became seriously interested in submarines. His name was Wilhelm Bauer. He was born in 1822 in a small town in Bavaria and, though a turner by trade, joined the army in 1842. Bauer was even in his youth of a highly inventive turn of mind. He possessed an indomitable will and an unlimited supply of enthusiasm. Step by step he acquired, in what little time he could spare from his military duties, the necessary mechanical knowledge, and finally, supported financially by a few loyal friends and patrons, he built his first submarine at Kiel at a cost of about $2750. It sank to the bottom on its first trial trip, fortunately without anyone on board. Undaunted he continued his efforts.

When he found that his support at Kiel was weakening, he promptly went to Austria. In spite of glowing promises, opposition on the part of some officials deprived Bauer of the promised assistance. He went then to England and succeeded in enlisting the interest of the Prince Consort. A boat was built according to Bauer's plans, which, however, he was forced by the interference of politicians to change to such an extent that it sank on its first trial with considerable loss of life.

Still full of faith in his ability to produce a successful submarine, Bauer now went to Russia. In 1855, he built a boat at St. Petersburg and had it accepted by the Russian Government. It was called *Le Diable Marin* and looked very much like a dolphin. Its length was fifty-

A British seaplane with folding wings

two feet, its beam twelve feet five inches, and its depth eleven feet. Its hull was of iron. A propeller, worked by four wheels, furnished motive power. Submersion and stability were regulated by four cylinders into which water could be pumped at will.

The first trial of the boat was made on May 26, 1856, and was entirely successful. In later trials as many as fourteen men at a time descended in *Le Diable Marin*. It is said that Bauer made a total of 134 trips on his boat. All but two were carried out successfully. At one time, however, the propeller was caught in some seaweed and it was only by the quickest action that all the water was pumped out and the bow of the boat allowed to rise out of the water, so that the occupants managed to escape by means of the hatchway. Like Fulton in France, Bauer now experienced in Russia a sudden decrease of official interest. When he finally lost his boat, about four weeks later, he also lost his courage, and in 1858 he returned to Germany where he later died in comparative poverty.

Contemporary with Bauer's submarines and immediately following them were a large number of other boats. Some of these were little more than freaks. Others failed in certain respects but added new features to the sum-total of submarine inventions. As early as 1854, M. Marié-Davy, Professor of Chemistry at Montpellier University, suggested an electro-magnetic engine as motive power. In 1855 a well-known engineer, J. Nasmith, suggested a submerged motor, driven by a steam engine. None of the boats of this period proved successful enough, however, to receive more than passing notice, and very few, indeed, ever reached the trial stage. But before long the rapid development of internal-combustion engines and the immense progress made in the study of electricity was to advance the development of submarines by leaps and bounds.

CHAPTER 12

The Coming of Steam and Electricity

In the fall of 1863, the Federal fleet was blockading the harbour of Charleston, S. C. Included among the many ships was one of the marvels of that period, the United States battleship *Ironsides*. Armour-plated and possessing what was then considered a wonderful equipment of high calibred guns and a remarkably trained crew, she was the terror of the Confederates. None of their ships could hope to compete with her and the land batteries of the Southern harbour were powerless to reach her.

During the night of October 5, 1863, the officer of the watch on board the *Ironsides*, Ensign Howard, suddenly observed a small object looking somewhat like a pleasure boat, floating close to his own ship. Before Ensign Howard's order to fire at it could be executed, the *Ironsides* was shaken from bow to stern, an immense column of water was thrown up and flooded her deck and engine room, and Ensign Howard fell, mortally wounded. The little floating object was responsible for all this. It was a Confederate submersible boat, only fifty feet long and nine feet in diameter, carrying a fifteen-foot spar-torpedo. She had been named *David* and the Confederate authorities hoped to do away by means of her with the Goliaths of the Federal navy. Manned only by five men, under the command of Lieutenant W. T. Glassel, driven by a small engine and propeller, she had managed to come up unobserved within striking distance of the big battleship.

The attack, however, was unsuccessful. The *Ironsides* was undamaged. On the other hand the plucky little *David* had been disabled to such an extent that her crew had to abandon her and take to the water, allowing their boat to drift without motive power. Four of them were later picked up. According to an account in Barnes, *Torpedoes and Torpedo Warfare*, the engineer, after having been in the water for some

A British anti-aircraft gun

time, found himself near her and succeeded in getting on board. He relighted her fires and navigated his little boat safely back to Charleston. There she remained, making occasional unsuccessful sallies against the Federal fleet, and when Charleston was finally occupied by the Federal forces, she was found there.

In spite of this failure the Confederates continued their attempts to break the blockade of their most important port by submarine devices. A new and somewhat improved *David* was ordered and built at another port. News of this somehow reached the Federal Navy Department and was immediately communicated to Vice-Admiral Dahlgren, in command of the blockading fleet. Despite this warning and instructions to all the officers of the fleet, the second *David* succeeded in crossing Charleston bar.

This new boat was a real diving submarine boat and though frequently called *David* had been christened the *Hundley*. It had been built in the shipyards of McClintock & Hundley at Mobile, Alabama, and had been brought to Charleston by rail. On her trial she proved very clumsy and difficult to manage. For her first trip a crew of nine men volunteered. Not having any conning tower it was necessary that one of the hatchways should be left open while the boat travelled on the surface so that the steersman could find his bearings. While she was on her first trip, the swell from a passing boat engulfed her. Before the hatchway could be closed, she filled with water. Of course, she sank like a piece of lead and her entire crew, with the exception of the steersman, was drowned.

In spite of this mishap the *Hundley* was raised and again put in commission. Lieutenant Payne who had steered her on her first fatal trip had lost neither his courage nor faith and again assumed command of her. Soon after she started on her second trip a sudden squall arose. Before the hatchways could be closed, she again filled with water and sank, drowning all of her crew with the exception of Lieutenant Payne and two of his men.

Undaunted he took her out on a third trip after she had again been raised. Ill luck still pursued her. Off Fort Sumter she was capsized and this time four of her crew were drowned.

The difficulties encountered in sailing the *Hundley* on the surface of the water apparently made no difference when it came to finding new crews for her. By this time, however, the powers that be had become anxious that their submarine boat should accomplish something against an enemy, instead of drowning only her own men and it was decided to use her on the next trip in a submerged state. Again Lieutenant Payne was entrusted with her guidance. Her hatches were closed, her water tanks filled, and she was off for her first dive. Something went wrong however; either too much water had been put in her tanks or else the steering gear refused to work. At any rate she hit the muddy bottom with such force that her nose became deeply imbedded and before she could work herself free her entire crew of eight was suffocated. Lieutenant Payne himself lost his life which he had risked so valiantly and frequently before.

Once more she was raised and once more volunteers rushed to man her. On the fifth trip, however, the *Hundley*, while travelling underwater, became entangled in the anchor chains of a boat she passed and was held fast so long that her crew of nine were dead when she was finally disentangled and raised.

Thirty-five lives had so far been lost without any actual results having been accomplished. In spite of this a new crew was found. Her commander, Lieutenant Dixon, was ordered to make an attack against the Federal fleet immediately, using, however, the boat as a submersible instead of a submarine.

Admiral David Porter in his *Naval History of the Civil War* described the attack, which was directed against the U. S. S. *Housatonic*, one of the newest Federal battleships, as follows:

> At about 8.45 P. M., the officer of the deck on board the unfortunate vessel discovered something about one hundred yards away, moving along the water. It came directly towards the ship, and within two minutes of the time it was first sighted was alongside. The cable was slipped, the engines backed, and all hands called to quarters. But it was too late—the torpedo struck the *Housatonic* just forward of the mainmast, on the starboard side, on a line with the magazine. The man who steered her (the *Hundley*) knew where the vital spots of the steamer were and he did his work well. When the explosion took place the ship trembled all over as if by the shock of an earthquake, and seemed to be lifted out of the water, and then sank stern foremost, heeling to port as she went down.

Only a part of the *Housatonic*'s complement was saved. Of the *Hundley* no trace was discovered and she was believed to have escaped. Three years later, however, divers who had been sent down to examine the hull of the *Housatonic* found the little submarine stuck in the hole made by her attack on the larger ship and inside of her the bodies of her entire crew.

The submarines and near-submarines built in the United States during the Civil War were remarkable rather for what they actually accomplished than for what they contributed towards the development of submarine boats. Perhaps the greatest service which they rendered in the latter direction was that they proved to the satisfaction of many scientific men that submarine boats really held vast possibilities as instruments of naval warfare.

France still retained its lead in furnishing new submarine projects. One of these put forward in 1861 by Olivier Riou deserves mention because it provided for two boats, one driven by steam and one by electricity. Both of these submarines were built, but inasmuch as nothing is known of the result of their trials, it is safe to conclude that neither of them proved of any practical value.

Two years later, in 1863, two other Frenchmen, Captain Bourgeois and M. Brun, built at Rochefort a submarine 146 feet long and 12 feet in diameter which they called the *Plongeur*. They fitted it with a compressed-air engine of eighty horse-power. Extensive trials were made with this boat but resulted only in the discovery that, though it was possible to sink or rise with a boat of this type without great difficulty, it was impossible to keep her at an even keel for any length of time.

During the next few years, undoubtedly as a result of the submarine activities during the Civil War, a number of projects were put forward in the United States, none of which, however, turned out successfully. One of them, for which a man by the name of Halstead was responsible, was a submarine built for the United States Navy in 1865. It was not tried out until 1872 and it was not even successful in living up to its wonderful name, *The Intelligent Whale*. Its first trial almost resulted in loss of life and was never repeated. In spite of this, however, the boat was preserved and may still be seen at the Brooklyn Navy Yard.

In the meantime, an invention had been made by an Austrian artillery officer which before long was to exert a powerful influence on submarine development, though it was in no sense a submarine boat. The manner in which the submarines had attacked their opponents during the Civil War suggested to him the need of improvements in this direction. As a result he conceived a small launch which was to carry the explosive without any navigators. Before he could carry his plans very far he died. A brother officer in the navy continued his work and finally interested the manager of an English engineering firm located at Fiume, Mr. Whitehead. The result of the collaboration of these two men was the Whitehead torpedo. A series of experiments led to the construction of what was first called a "Submarine Locomotive" torpedo, which not only contained a sufficient quantity of explosives to destroy large boats, but was also enabled by mechanical means to propel itself and keep on its course after having been fired. The Austrian Government was the first one to adopt this new weapon. Whitehead, however, refused to grant a monopoly to the Austrians and in 1870 he sold his manufacturing rights and secret processes to the British Government for a consideration of $45,000.

Before very long, special boats were built for the purpose of carrying and firing these torpedoes and gradually every great power developed a separate torpedo flotilla. Hand in hand with this development a large number of improvements were made on the original torpedo and some of these devices proved of great usefulness in the development of submarine boats.

The public interest in submarines grew rapidly at this time. Every man who was a boy in 1873, or who had the spirit of boyhood in him then,—or perhaps now,—will remember the extraordinary piece of literary and imaginative prophecy achieved by Jules Verne in his novel *Twenty Thousand Leagues Under the Sea*. Little about the *Nautilus* that held all readers entranced throughout his story is lacking in the submarines of to-day except indeed its extreme comfort, even luxury. With those qualities our submarine navigators have to dispense. But the electric light, as we know it, was unknown in Verne's time yet he installed it in the boat of his fancy. Our modern internal-combustion engines were barely dreamed of, yet they drove his boat. His fancy even enabled him to foresee one of the most amazing features of the Lake boat of to-day, namely the compressed air chamber which opened to the sea still holds the water back, and enables the submarine navigator clad in a diver's suit to step into the wall of water and prosecute his labours on the bed of the ocean. Jules Verne even foresaw the callous and inhuman character of the men who command the German submarines to-day. His Captain Nemo had taken a vow of hate against the world and relentlessly drove the prow of his steel boat into the hulls of crowded passenger ships, finding his greatest joy in sinking slowly beside them with the bright glare of his submarine electric lights turned full upon the hapless women and children over whose sufferings he gloated as they sank. The man who sank the *Lusitania* could do no more.

More and more determined became the attempts to build submarine boats that could sink and rise easily, navigate safely and quickly, and sustain human beings under the surface of the water for a considerable length of time. Steam, compressed air, and electricity were called upon to do their share in accomplishing this desired result. Engineers in every part of the world began to interest themselves in the submarine problem and as a result submarine boats in numbers were either projected or built between 1875 and 1900.

One of the most persistent workers in this period was a well-known Swedish inventor, Nordenfeldt, who had established for himself a reputation by inventing a gun which even to-day has lost nothing of its fame. In 1881 he became interested in the work which had been done by an English clergyman named Garret. The latter had built a submarine boat which he called the *Resurgam* (I shall rise)—thus neatly combining a sacred promise with a profane purpose. In 1879 another boat was built by him driven by a steam engine. Nordenfeldt used the fundamental ideas upon which these two boats were based, added to them some improvements of his own as well as some devices which had been used by Bushnell, and finally launched in 1886 his first submarine boat. The government of Greece bought it after some successful trials. Not to be outdone, Greece's old rival, Turkey, immediately ordered two boats for her own navy. Both of these were much larger than the Greek boat and by 1887 they had reached Constantinople in sections where they were to be put together. Only one of them, however, was ever completed. Characteristic Turkish delay intervened. The most typical feature of this boat was the fact that it carried a torpedo tube for Whitehead torpedoes. On the surface of the water this boat proved very efficient, but as an underwater boat it was a dismal failure. More than in any other craft that had ever been built and accepted, the lack of stability was a cause of trouble in the *Nordenfeldt II*. As soon as any member of the crew moved from one part of the boat to another, she would dip in the direction in which he was moving, and everybody, who could not in time take hold of some part of the boat, came sliding and rolling in the same direction. When finally such a tangle was straightened out, only a few minutes elapsed before somebody else, moving a few steps, would bring about the same deplorable state of affairs. The *Nordenfeldt II.* acted more like a bucking bronco than a self-respecting submarine boat and as a result it became impossible to find a crew willing to risk their lives in manning her. Before very long she had rusted and rotted to pieces. In spite

An anti-aircraft outpost

of this lack of success, Nordenfeldt built a fourth boat which displayed almost as many unfortunate features as her predecessors and soon was discarded and forgotten.

In the latter part of the nineteenth century the French Government, which for so many years had shown a strong and continuous interest in the submarine problem, was particularly active. Three different types of boats built in this period under the auspices and with the assistance of the French Government deserve particular attention. The first of these was the *Gymnote*, planned originally by a well-known French engineer, Dupuy de Lome, whose alert mind also planned an airship and made him a figure in the history of our Panama Canal. He died, however, before his project could be executed. M. Gustave Zédé, a marine engineer and his friend, continued his work after modifying some of his plans. The French Minister of Marine of this period, Admiral Aube who had long been strongly interested in submarines, immediately accepted M. Zédé's design and ordered the boat to be built. As the earliest of successful submarines she merits description:

The *Gymnote* was built of steel in the shape of a cigar. She was 59 feet long, 5 feet 9 inches beam, and 6 feet in diameter, just deep enough to allow a man to stand upright in the interior. The motive power was originally an electro-motor of 55 horse-power, driven from 564 accumulators. It was of extraordinary lightness, weighing only 4410 pounds, and drove the screw at the rate of two thousand revolutions a minute, giving a speed of six knots an hour, its radius of action at this speed being thirty-five miles.

Immersion was accomplished by the introduction of water into three reservoirs, placed one forward, one aft, and one centre. The water was expelled either by means of compressed air or by a rotary pump worked by an electro-motor. Two horizontal rudders steered the boat in the vertical plane and an ordinary rudder steered in the horizontal.

The *Gymnote* had her first trial on September 4, 1888, and the Paris *Temps* described the result in the following enthusiastic language:

> She steered like a fish both as regards direction and depth; she mastered the desired depth with ease and exactness; at full power she attained the anticipated speed of from nine to ten knots; the lighting was excellent, there was no difficulty about heating. It was a strange sight to see the vessel skimming along the top of the water, suddenly give a downward plunge with its snout,

A COAST DEFENCE ANTI-AIRCRAFT GUN

and disappear with a shark-like wriggle of its stern, only to come up again at a distance out and in an unlooked-for direction. A few small matters connected with the accumulators had to be seen to, but they did not take a month.

Following along the same lines as this boat another boat, considerably larger, was built. Before it was completed, M. Zédé died and it was decided to name the new boat in his honour. The *Gustave Zédé* was launched at Toulon on June 1, 1893; she was 159 feet in length, beam 12 feet 4 inches, and had a total displacement of 266 tons. Her shell was of "Roma" bronze, a non-magnetic metal, and one that could not be attacked by sea water.

The motive power was furnished by two independent electro-motors of 360 horse-power each and fed by accumulators. In order to endow the boat with a wide radius of action a storage battery was provided.

The successive crews of the *Gustave Zédé* suffered much from the poisonous fumes of the accumulators, and during the earlier trials all the men on board were ill.

In the bows was a torpedo tube, and an arrangement was used whereby the water that entered the tube after the discharge of the torpedo was forced out by compressed air. Three Whitehead torpedoes were carried. In spite of the fact that a horizontal rudder placed at the stern had not proved serviceable on the *Gymnote*, such a rudder was fitted in the *Gustave Zédé*. With this rudder she usually plunged at an angle of about 5°, but on several occasions she behaved in a very erratic fashion, seesawing up and down, and once when the Committee of Experts were on board, she proved so capricious, going down at an angle of 30°-35°, often throwing the poor gentlemen on to the floor, that it was decided to fix a system of six rudders, three on each side.

Four water tanks were carried, one at each end and two in the middle, and the water was expelled by four pumps worked by a little electro-motor; these pumps also furnished the air necessary for the crew and for the discharge of the torpedoes. For underwater vision, an optical tube and a periscope had been provided.

On July 5, 1899, still another submarine boat was launched for the French Navy. She was called the *Morse*. She was 118 feet long, 9 feet beam, displaced 146 tons, and was likewise made of "Roma" bronze. The motive power was electricity and in many other respects she was very similar to the *Gustave Zédé*, embodying, however, a number of improvements. M. Calmette, who accompanied the French Minister of War on the trial trip of the *Morse*, described his experience in the Paris *Figaro* as follows:

> General André, Dr. Vincent, a naval doctor, and I entered the submarine boat *Morse* through the narrow opening in the upper surface of the boat. Our excursion was to begin immediately; in two hours we came to the surface of the water again three miles to the north to rejoin the *Narval*. Turning to the crew, every man of which was at his post, the commandant gave his orders, dwelling with emphasis on each word. A sailor repeated his orders one by one, and all was silent. The *Morse* had already started on its mysterious voyage, but was skimming along the surface until outside the port in order to avoid the numerous craft in the Arsenal. To say that at this moment, which I had so keenly anticipated, I did not have the tremor which comes from contact with the unknown would be beside the truth. On the other hand, calm and imperturbable, but keenly curious as to this novel form of navigation, General André had already taken his place near the commandant on a folding seat. There were no chairs in this long tube in which we were imprisoned. Everything was arranged for the crew alone, with an eye to serious action. Moreover, the Minister of War was too tall to stand upright beneath the iron ceiling, and in any case it would be impossible to walk about.
>
> The only free space was a narrow passage, sixty centimetres broad, less than two metres high, and thirty metres long, divided into three equal sections. In the first, in the forefront of the tube, reposed the torpedoes, with the machine for launching them, which at a distance of from 500 to 600 metres were bound to sink, with the present secret processes, the largest of ironclads. In the second section were the electric accumulators which gave the light and power. In the third, near the screw, was the electric motor which transformed into movement the current of the accumulators. Under all this, beneath the floor, from end to end, were immense water ballasts, which were capable of being emptied or filled in a few seconds by electric machines, in order to carry the vessel up or down. Finally, in the centre of the tube, dominating these three sections, which the electric light inundated, and which no partition divided, the navigating lieutenant stood on the lookout giving his orders.

THE SUBMARINE'S PERFECT WORK

There was but one thing which could destroy in a second all the sources of authority, initiative, and responsibility in this officer. That was the failure of the accumulators. Were the electricity to fail everything would come to a stop. Darkness would overtake the boat and imprison it for ever in the water. To avoid any such disaster there have been arranged, it is true, outside the tube and low down, a series of lead blades which were capable of being removed from within to lighten the vessel. But admitting that the plunger would return to the surface, the boat would float hither and thither, and at all events lose all its properties as a submarine vessel. To avoid any such disaster a combination of motors have been in course of construction for some months, so that the accumulators might be loaded afresh on the spot, in case of their being used up.

The *Morse*, after skimming along the surface of the water until outside the port, was now about to sink. The commandant's place was no longer in the helmet or kiosk whence he could direct the route along the surface of the sea. His place was henceforth in the very centre of the tube, in the midst of all sort of electric manipulators, his eyes continually fixed on a mysterious optical apparatus, the periscope. The other extremity of this instrument floated on the surface of the water, and whatever the depth of the plunge it gave him a perfectly faithful and clear representation, as in a camera, of everything occurring on the water.

The most interesting moment of all now came. I hastened to the little opening to get the impression of total immersion. The lieutenant by the marine chart verified the depths. The casks of water were filled and our supply of air was thereby renewed from their stores of surplus air. In our tiny observatory, where General André stationed himself above me, a most unexpected spectacle presented itself as the boat was immersed.

The plunge was so gentle that in the perfect silence of the waters one did not perceive the process of descent, and there was only an instrument capable of indicating, by a needle, the depth to which the *Morse* was penetrating. The vessel was advancing while at the same time it descended, but there was no sensation of either advance or roll. As to respiration, it was as perfect as in any room. M. de Lanessan, who since entering office has ordered eight more submarine vessels, had concerned himself with the question as a medical man also, and, thanks to the labours of a commission formed by him, the difficulties of respiration were entirely solved. The crew were able to remain under water sixteen hours without the slightest strain. Our excursion on this occasion lasted scarcely two hours. Towards noon, by means of the mysterious periscope, which, always invisible, floated on the surface and brought to the vessel below a reflection of all that passed up above, the captain showed us the *Narval*, which had just emerged with its two flags near the old battery *Impregnable*. From the depths in which we were sailing we watched its slightest manoeuvres until the admiral's flag, waving on the top of a fort, reminded us that it was time to return.

CHAPTER 13
John P. Holland and Simon Lake

The Naval Committee of the House of Representatives of the United States in the early part of 1900 held a meeting for the purpose of hearing expert testimony upon the subject of submarines. Up to then the United States authorities had shown, as compared with the ruling powers of other navies, only a limited amount of interest in the submarine question. Increased appropriations for the construction of submarine boats which were then beginning to become more frequent in other countries acted, however, as a stimulus at this time.

The committee meeting took place a few days after some of the members of the committee, together with a number of United States navy officers, had attended an exhibition of a new submarine boat, the *Holland No. 9*.

The late Admiral Dewey gave the following opinion about this submarine to the committee, an opinion which since then has become rather famous:

> Gentlemen: I saw the operation of the boat down off Mount Vernon the other day. Several members of this committee were there. I think we were very much impressed with its performance. My aid, Lieutenant Caldwell, was on board. The boat did everything that the owners proposed to do. I said then, and I have said it since, that if they had two of those things at Manila, I could never have held it with the squadron I had. The moral effect—to my mind, it is infinitely superior to mines or torpedoes or anything of the kind. With two of those in Galveston all the navies of the world could not blockade the place.

Admiral Dewey's approval of the *Holland No. 9* undoubtedly exerted a considerable influence on the Naval Committee and as a result of

Types of American aircraft

its recommendations the United States Government finally purchased the boat on April 11, 1900, for $150,000. This amount was about $86,000 less than the cost of building to the manufacturers, the Holland Torpedo Boat Company. The latter, however, could well afford to take this loss because this first sale resulted a few months afterwards—on August 25th—in an order for six additional submarines. The British Government also contracted in the fall of the same year for five Hollands. The navy of almost every power interested in submarines soon followed the lead of the British Admiralty. Submarines of the Holland type were either ordered outright, or else arrangements were concluded permitting the use of the basic patents held by the Holland Company. It will be noted that the United States Government having discovered that it had a good thing benevolently shared it with the governments that might be expected to use it against us.

The *Holland No. 9*, as her very name indicates, was one of a long line of similar boats. As compared with other experimental submarine boats she was small. She was only fifty-three feet ten inches long, and ten feet seven inches deep. Although these proportions made her look rather thickset, they were the result of experimental work done by the builder during a period of twenty-five years. She was equipped both with a gasoline engine of fifty horse-power and an electric motor run by storage batteries. The latter was intended for use when the boat was submerged, the former when she was travelling on the surface of the water. She was capable of a maximum speed of seven knots an hour. Her cruising radius was 1500 miles and the combination of oil and electric motors proved so successful that from that time on every submarine built anywhere adopted this principle. Two horizontal rudders placed at the stern of the boat steered her downward whenever she wanted to dive and so accomplished a diver was this boat that a depth of twenty-eight feet could be reached by her in five seconds. Her conning tower was the only means of making observations. No periscopes had been provided because none of the instruments available at that time gave satisfaction. This meant that whenever she wished to aim at her target it was necessary for her to make a quick ascent to the surface. Her stability was one of her most satisfactory features. So carefully had her proportions been worked out that there was practically no pitching or rolling when the boat was submerged. Even the concussion caused by the discharge of a torpedo was hardly noticeable because arrangements had been made to take up the recoil caused by the firing and to maintain the balance of the boat by permitting a quantity of water equal to the weight of the discharged torpedo to enter special compartments at the very moment of the discharge.

The *Holland No. 9* was built at Lewis Nixon's shipyards at Elizabethport, New Jersey, and was launched early in 1898, just previous to the outbreak of the Spanish-American War. Although numerous requests were made to the United States Government by her inventor and builder, John P. Holland, for permission to take her into Santiago harbour in an attempt to torpedo Cervera's fleet, the navy authorities at Washington refused this permission. Why? Presumably through navy hostility to the submarine idea. When the *Monitor* whipped the *Merrimac* in 1862 the former ship belonged to her inventor, not to the United States Government. It would have been interesting had Holland at his own expense destroyed the Spanish ships.

John P. Holland at the time when he achieved his success was fifty-eight years old, Irish by birth and an early immigrant to the United States. He had been deeply interested for many years in mechanical problems and especially in those connected with navigation. The change from the old wooden battleships to the new ironclads and the rapidly increasing development of steam-engines acted as a strong stimulus to the young Irishman's experiments. It is claimed that his interest in submarine navigation was due primarily to his desire to find a weapon strong enough to destroy or at least dominate the British navy; for at that time Holland was strongly anti-British, because he, like many other educated Irishmen of that period, desired before everything else to free Ireland. His plans for doing this by supplying to the proposed Irish Republic a means for overcoming the British navy found little support and a great deal of ridicule on the part of his Irish friends. In spite of this he kept on with his work and in 1875 he built and launched his first submarine boat at Paterson. This boat was far from being very revolutionary. She was only sixteen feet long and two feet in diameter, shaped like a cigar but with both ends sharply pointed. In many respects except in appearance she was similar to Bushnell's *Turtle*. Room for only one operator was provided and the latter was to turn the propeller by means of pedals to be worked by his feet. She accomplished little beyond giving an opportunity to her inventor and builder to gather experience in actual underwater navigation.

Two years later in 1877 the *Holland No. 2* was built. In spite of the number of improvements represented by her she was not particularly successful. Her double hull, it is true, provided space for carrying water ballast. But the leaks from this ballast tank continuously threatened to drown the navigator sitting inside of the second hull. A small oil engine of four horse-power was soon discarded on account of its inefficiency.

The experience gathered by Holland in building and navigating these two boats strengthened his determination to build a thoroughly successful submarine and increased his faith in his ability to do so. He opened negotiations with the Fenian Brotherhood. This was a secret society founded for the purpose of freeing Ireland from British rule and creating an Irish Republic. Holland finally succeeded in persuading his Fenian friends to order from him two submarine boats and to supply him with the necessary means to build them. Both of these boats were built. The lack of success of the first one was due primarily to the inefficiency of her engine. The second boat which was really the *Holland No. 4* was built in 1881. It is usually known as the *Fenian Ram*, and is still in existence at New Haven, Connecticut, where a series of financial and political complications finally landed her.

These two boats added vastly to Holland's knowledge concerning submarine navigation. A few others which he built with his own means increased this fund of knowledge and step by step he came nearer to his goal. By 1888 his reputation as a submarine engineer and navigator had grown to such an extent that Holland was asked by the famous Philadelphia shipbuilders, the Cramps, to submit to them designs for a submarine boat to be built by the United States Government. Only one other design was submitted and this was by the Scandinavian, Nordenfeldt.

William C. Whitney, then Secretary of the United States Navy, accepted Holland's design. Month after month passed by wasted by the usual governmental red tape, and when all preliminary arrangements had been made and the contract for the actual building of an experimental boat was to be drawn up, a sudden change in the administration resulted in the dropping of the entire plan.

Holland's faith in the future submarine and in his own ability was still unshaken, but this was not the case with his financial condition. None of the boats he had built so far had brought him any profits and on some he had lost everything that he had put into them. His financial support, for which he relied entirely upon relatives and friends, was practically exhausted. But fortunately on March 3, 1893, Congress appropriated a sum of money to defray the expenses of constructing an experimental submarine. Invitations to inventors were extended.

So precarious was Holland's financial condition at that time that he found it necessary to borrow the small sum of money involved in making plans which he had to submit. It is claimed that he succeeded in doing this in a manner highly typical of his thoroughness.

He needed only about $350.00 but even this comparatively small sum was more than he had. However, he happened to be lunching with a young lawyer just about this time and began to tell him about his financial difficulties. Holland told him that if he only had $347.19 he could prepare the plans and pay the necessary fees. And that done, he was sure of being able to win the competition. His lawyer friend, of course, had been approached before by other people for loans. Invariably they had asked him for some round sum and Holland's request for $347.19 when he might just as well have asked for $350.00 aroused his interest. He asked the inventor what the nineteen cents were to be used for. Quick as a flash he was told that they were needed to pay for a particular type of ruler necessary to draw the required plans. So impressed was the lawyer with Holland's accuracy and honesty in asking not a cent more than he actually needed that he at once advanced the money. And a good investment it turned out to be. For in exchange he received a good-sized block of stock in the Holland Torpedo Boat Company which in later years made him a multi-millionaire.

Holland's plans did win the competition just as he asserted that they would; but, of course, winning a prize, offered by a government, and getting that government to do something about it, are two different matters. So two years went by before the Holland Torpedo Boat Company at last was able to start with the construction of the new submarine which was to be called the *Plunger*.

The principal feature of this new boat was that it was to have a steam engine for surface navigation and an electric motor for underwater navigation. This arrangement was not so much a new invention of Holland's as an adaptation of ideas which had been promulgated by others. Especially indebted was he in this respect to Commander Hovgaard of the Danish navy who, in 1887, had published an important book on the subject of double propulsion in submarines. Though Holland had made many improvements on these earlier theories, he soon found out that even at that there was going to be serious trouble with the *Plunger*'s engines. The boat had been launched in 1897; but instead of finishing it, he persuaded the government to permit his company to build a new boat, and to return to the government all the money so far expended on the *Plunger*.

The new boat, *Holland No. 8*, was started immediately and completed in record time but she, too, was unsatisfactory to the inventor. So without loss of time he went ahead and built another boat, the *Holland No. 9*, which, as we have said, became the first United States submarine.

Two other men submitted plans for submarine boats in the competition which was won by the Holland boat, George C. Baker and Simon Lake. Neither of these was accepted. Mr. Baker made no further efforts to find out if his plans would result in a practicable submarine boat. But Simon Lake was not so easily discouraged.

It is very interesting that the United States Navy Department at that time demanded that plans submitted for this competition should meet the following specifications:

1. Safety.
2. Facility and certainty of action when submerged.
3. Speed when running on the surface.
4. Speed when submerged.
5. Endurance, both submerged and on the surface.
6. Stability.
7. Visibility of object to be attacked.

In spite of the many years that have passed since this competition and in spite of the tremendous progress that has been made in submarine construction these are still the essential requirements necessary to make a successful submarine boat.

The designs submitted by Mr. Lake provided for a twin-screw vessel, 80 feet long, 10 feet beam, and 115 tons displacement, with 400 horse-power steam engines for surface propulsion and 70 horse-power motors for submerged work. The boat was to have a double hull, the spaces between the inner and the outer hulls forming water ballast tanks. There were to be four torpedo tubes, two forward and two aft.

In an article published in 1915 in *International Marine Engineering*, Mr. Lake says about his 1893 design:

> The new and novel feature which attracted the most attention and scepticism regarding this design was (the author was later informed by a member of the board) the claim made that the vessel could readily navigate over the waterbed itself, and that while navigating on the waterbed a door could be opened in the bottom of a compartment and the water kept from entering the vessel by means of compressed air, and that the crew could, by donning diving suits, readily leave and enter the vessel while sub-

merged. Another novel feature was in the method of controlling the depth of submergence when navigating between the surface and waterbed. The vessel was designed to always submerge and navigate on a level keel rather than to be inclined down or up by the back, to "dive" or "rise." This maintenance of a level keel while submerged was provided for by the installation of four depth regulating vanes which I later termed "hydroplanes" to distinguish them from the forward and aft levelling vanes or horizontal rudders. These hydroplanes were located at equal distances forward and aft of the centre of gravity and buoyancy of the vessel when in the submerged condition, so as not to disturb the vessel when the planes were inclined down or up to cause the vessel to submerge or rise when under way.

I also used, in conjunction with the hydroplanes, horizontal rudders which I then called "levelling vanes," as their purpose was just the opposite from that of the horizontal rudder used in the diving type of vessel. They were operated by a pendulum controlling device to be inclined so as to always maintain the vessel on a level keel rather than to cause her to depart therefrom. When I came to try this combination out in practice, I found hand control of the horizontal rudders was sufficient. If vessels with this system of control have a sufficient amount of stability, you will run for hours and automatically maintain both a constant depth and a level keel, without the depth control man touching either the hydroplane or horizontal rudder control gear. This automatic maintenance of depth without manipulating the hydroplanes or rudders was a performance not anticipated, nor claimed in my original patent on the above-mentioned combination, and what caused these vessels to function in this manner remained a mystery, which was unsolved until I built a model tank in 1905 in Berlin, Germany, and conducted a series of experiments on models of submarines. I then learned that a down pull of a hydroplane at a given degree of inclination varied according to its depth of submergence and that the deeper the submergence, the less the down pull. This works out to give automatic trim on a substantially level keel, and I have known of vessels running for a period of two hours without variation of depth of one foot and without once changing the inclination of either the hydroplanes or the horizontal rudder.

A great deal of scepticism was displayed for many years towards this new system of controlling the depth of submergence. But in recent years all the latest submarine boats have been built on this plan.

Who, then, was this mechanical genius who was responsible for these far-going changes in submarine construction? Simon Lake was born at Pleasantville, New Jersey, September 4, 1866. He was educated at Clinton Liberal Institute, Fort Plain, New York, and Franklin Institute, Philadelphia. Early in life he displayed a marked interest in and genius for mechanical problems. His lack of success in the 1893 competition only spurred him on to further efforts. As long as the United States Government was unwilling to assist him in building his submarine boat, there was nothing left for him except to build it from his own means. In 1894, therefore, he set to work on an experimental boat, called the *Argonaut, Jr.* According to Mr. Lake's description as published in *International Marine Engineering* in a series of articles from his pen the *Argonaut, Jr.*, was—

> .... provided with three wheels, two on either side forward and one aft, the latter acting as a steering wheel. When on the bottom the wheels were rotated by hand by one or two men inside the boat. Her displacement was about seven tons, yet she could be propelled at a moderate walking gait when on the bottom. She was also fitted with an air lock and diver's compartment, so arranged that by putting an air pressure on the diver's compartment equal to the water pressure outside, a bottom door could be opened and no water would come into the vessel. Then by putting on a pair of rubber boots the operator could walk around on the sea bottom and push the boat along with him and pick up objects, such as clams, oysters, etc. from the sea bottom.

So much interest was aroused by this little wooden boat that Mr. Lake was enabled to finance the building of a larger boat, called the *Argonaut*. It was designed in 1895 and built in 1897 at Baltimore.

Concerning the *Argonaut* Mr. Lake says in the same article:

> The *Argonaut* as originally built was 36 feet long and 9 feet in diameter. She was the first submarine to be fitted with an internal-combustion engine. She was propelled with a thirty horse-power gasoline (petrol) engine driving a screw propeller. She was fitted with two toothed driving wheels forward which were revolved by suitable gearing when navigating on the waterbed, or they could be disconnected from this gearing

and permitted to revolve freely, propulsion being secured by the screw propeller. A wheel in the rudder enabled her to be steered in any direction when on the bottom. She also had a diving compartment to enable divers to leave or enter the vessel when submerged, to operate on wrecks or to permit inspection of the bottom or to recover shellfish. She also had a lookout compartment in the extreme bow, with a powerful searchlight to light up a pathway in front of her as she moved along over the waterbed. This searchlight I later found of little value except for night work in clear water. In clear water the sunlight would permit of as good vision without the use of the light as with it, while if the water was not clear, no amount of light would permit of vision through it for any considerable distance.

In January, 1898, while the *Argonaut* was submerged, telephone conversation was held from submerged stations with Baltimore, Washington, and New York.

In 1898, also, the *Argonaut* made the trip from Norfolk to New York under her own power and unescorted. In her original form she was a cigar-shaped craft with only a small percentage of reserve buoyancy in her surface cruising condition. We were caught out in the severe November northeast storm of 1898 in which over 200 vessels were lost and we did not succeed in reaching a harbour in the "horseshoe" back of Sandy Hook until, of course, in the morning. The seas were so rough they would break over her conning tower in such masses I was obliged to lash myself fast to prevent being swept overboard. It was freezing weather and I was soaked and covered with ice on reaching harbour.

This experience caused me to apply to the *Argonaut* a further improvement for which I had already applied for a patent. This was, doubled around the usual pressure resisting body of a submarine, a ship-shape form of light plating which would give greater seaworthiness, better surface speed, and make the vessel more habitable for surface navigation. It would, in other words, make a "sea-going submarine," which the usual form of cigar-shaped vessel was not, as it would not have sufficient surface buoyancy to enable it to rise with the seas and the seas would sweep over it as they would sweep over a partly submerged rock.

The *Argonaut* was, therefore, taken to Brooklyn, twenty feet added to her length, and a light water-tight buoyancy superstructure of ship-shape form added. This superstructure was opened to the sea when it was desired to submerge the vessel, and water was permitted to enter the space between the light plating of the ship-shaped form and the heavy plating of the pressure resisting hull. This equalized pressure on the light plates and prevented their becoming deformed due to pressure. The superstructure increased her reserve of buoyancy in the surface cruising condition from about 10 per cent. to over 40 per cent. and lifted right up to the seas like any ordinary type of surface vessel, instead of being buried by them in rough weather.

This feature of construction has been adopted by the Germans, Italians, Russians, and in all the latest types of French boats. It is the principal feature which distinguishes them in their surface appearance from the earlier cigar-shaped boats of the diving type. This ship-shaped form of hull is only suited to the level keel submergence.

In those days submarine boats were a much more unusual sight than they are to-day and simple fishermen who had never read or heard about submarines undoubtedly experienced disturbing sensations when they ran across their first underwater boat. Mr. Lake, a short time ago, while addressing a meeting of electrical engineers in Brooklyn, told the following experience which he had on one of his trips in the *Argonaut*:

> On the first trip down the Chesapeake Bay, we had been running along in forty feet of water and had been down about four hours. Night was coming on, so we decided to come up to find out where we were. I noticed one of those Chesapeake "Bug Eyes" lighting just to leeward of us, and, as I opened the conning tower hatch, called to the men aboard to find out where we were. As soon as I did so, he turned his boat around and made straight for the beach. I thought he was rather discourteous. He ran his boat up on that beach and never stopped; the last I saw of him was when he jumped ashore and started to run inland as hard as he and his helper could go. Finally I learned we were just above the mouth of the York or Rappahannock River and I found a sort of inland harbour back of it. I decided to put up there for the night. Then learning that there was a store nearby, we called after dark for more provisions and I noticed a large crowd there. We got what we wanted, and stepped outside the door. He asked us where we were from. "We are down

here in the submarine boat, *Argonaut*, making an experimental trip down the bay." He then commenced to laugh. "That explains it," he said; "just before nightfall, Captain So-and-So and his mate came running up here to the store just as hard as they could, and both dropped down exhausted, and when we were able to get anything out of them, they told a very strange story. That's why all these people are here." This is the story the storekeeper told me: "The men were out dredging and all at once they noticed a buoy with a red flag on it, and that buoy was going against the tide, and they could not understand it. It came up alongside, and they heard a 'puff, puff,' something like a locomotive puffing, and then they smelt sulphur." (The "puff, puff" was the exhaust of our engine and those fumes were what they thought was sulphur.) "Just then the thing rose up out of the water, then the smokestack appeared, and then the devil came right out of that smokestack."

In the January, 1899, issue of *McClure's Magazine* there appeared a profusely illustrated article entitled "Voyaging under the Sea." The first part of it, "The Submarine Boat *Argonaut* and her Achievements," was written by Simon Lake himself. In it he quotes as follows from the log book of the *Argonaut* under date of July 28, 1898.

> Submerged at 8.20 a.m. in about thirty feet of water. Temperature in living compartment, eighty-three degrees Fahrenheit. Compass bearing west-north-west, one quarter west. Quite a lively sea running on the surface, also strong current. At 10.45 A.M. shut down engine; temperature, eighty-eight degrees Fahrenheit.
>
> After engine was shut down, we could hear the wind blowing past our pipes extending above the surface; we could also tell by the sound when any steamers were in the vicinity. We first allowed the boat to settle gradually to the bottom, with the tide running ebb; after a time the tide changed, and she would work slightly sideways; we admitted about four hundred pounds of water additional, but she still would move occasionally, so that a pendulum nine inches long would sway one eighth of an inch (thwartship). At 12 o'clock (noon) temperature was eighty-seven degrees Fahrenheit; at 2.45 p.m. the temperature was still eighty-seven degrees Fahrenheit. There were no signs of carbonic acid gas at 2.45, although the engine had been closed down for three hours and no fresh air had been

admitted during the time. Could hear the whistle of boats on the surface, and also their propellers when running close, to the boat. At 3.30 the temperature had dropped to eighty-five degrees. At 3.45 found a little sign of carbonic acid gas, very slight, however, as a candle would burn fairly bright in the pits. Thought we could detect a smell of gasoline by comparing the fresh air which came down the pipe (when hand blower was turned). Storage lamps were burning during the five hours of submergence, while engine was not running.

At 3.50 engine was again started, and went off nicely. Went into diving compartment and opened door; came out through air-lock, and left pressure there; found the wheels had buried about ten inches or one foot, as the bottom had several inches of mud. We had 500 pounds of air in the tanks, and it ran the pressure down to 250 pounds to open the door in about thirty feet.

The temperature fell in the diving compartment to eighty-two degrees after the compressed air was let in.

Cooked clam fritters and coffee for supper. The spirits of the crew appeared to improve the longer we remained below; the time was spent in catching clams, singing, trying to waltz, playing cards, and writing letters to wives and sweethearts.

Our only visitors during the day were a couple of black bass that came and looked in at the windows with a great deal of apparent interest.

In future boats, it will be well to provide a smoking compartment, as most of the crew had their smoking apparatus all ready as soon as we came up.

Started pumps at 6.20, and arrived at the surface at 6.30. Down altogether ten hours and fifteen minutes. People on pilot boat *Calvert* thought we were all hands drowned.

The second part of this article was called "A Voyage on the Bottom of the Sea." It was written by Ray Stannard Baker, who had been fortunate enough to receive an invitation from Mr. Lake to accompany him on one of the trips of the *Argonaut*. Any one who has read Jules Verne's fascinating story *Twenty Thousand Leagues Under the Sea* must be struck immediately with the similarity between Mr. Baker's experiences and those of Captain Nemo's guests. It is not at all surprising, therefore, to have Mr. Baker tell us that during this trip Mr. Lake told him:

"When I was ten years old, I read Jules Verne's *Twenty Thousand Leagues Under the Sea*, and I have been working on submarine boats ever since."

Mr. Baker's record of what he saw and how he felt is not only a credit to his keen powers of observation, but also a proof of the fact that, in many ways, there was little difference between the *Argonaut* of 1898 and the most up-to-date submarine of to-day. In part he says:

> Simon Lake planned an excursion on the bottom of the sea for October 12, 1898. His strange amphibian craft, the *Argonaut*, about which we had been hearing so many marvels, lay off the pier at Atlantic Highlands. Before we were near enough to make out her hulk, we saw a great black letter A, framed of heavy gas-pipe, rising forty feet above the water. A flag rippled from its summit. As we drew nearer, we discovered that there really wasn't any hulk to make out—only a small oblong deck shouldering deep in the water and supporting a slightly higher platform, from which rose what seemed to be a squatty funnel. A moment later we saw that the funnel was provided with a cap somewhat resembling a tall silk hat, the crown of which was represented by a brass binnacle. This cap was tilted back, and as we ran alongside, a man stuck his head up over the rim and sang out, "Ahoy there!"
>
> A considerable sea was running, but I observed that the *Argonaut* was planted as firmly in the water as a stone pillar, the big waves splitting over her without imparting any perceptible motion.
>
> We scrambled up on the little platform, and peered down through the open conning-tower, which we had taken for a funnel, into the depths of the ship below. Wilson had started his gasoline engine.
>
> Mr. Lake had taken his place at the wheel, and we were going ahead slowly, steering straight across the bay toward Sandy Hook and deeper water. The *Argonaut* makes about five knots an hour on the surface, but when she gets deep down on the sea bottom, where she belongs, she can spin along more rapidly.
>
> The *Argonaut* was slowly sinking under the water. We became momentarily more impressed with the extreme smallness of the craft to which we were trusting our lives. The little platform around the conning-tower on which we stood—in

FOR ANTI-AIRCRAFT SERVICE

reality the top of the gasoline tank—was scarcely a half dozen feet across, and the *Argonaut* herself was only thirty-six feet long. Her sides had already faded out of sight, but not before we had seen how solidly they were built—all of steel, riveted and reinforced, so that the wonder grew how such a tremendous weight, when submerged, could ever again be raised.

I think we made some inquiries about the safety of submarine boats in general. Other water compartments had been flooded, and we had settled so far down that the waves dashed repeatedly over the platform on which we stood—and the conning-tower was still wide open, inviting a sudden engulfing rush of water. "You mustn't confuse the *Argonaut* with ordinary submarine boats," said Mr. Lake. "She is quite different and much safer."

He explained that the *Argonaut* was not only a submarine boat, but much besides. She not only swims either on the surface or beneath it, but she adds to this accomplishment the extraordinary power of diving deep and rolling along the bottom of the sea on wheels. No machine ever before did that. Indeed, the Argonaut is more properly a "sea motorcycle" than a "boat." In its invention Mr. Lake elaborated an idea which the United States Patent Office has decided to be absolutely original.

We found ourselves in a long, narrow compartment, dimly illuminated by yellowish-green light from the little round, glass windows. The stern was filled with Wilson's gasoline engine and the electric motor, and in front of us toward the bow we could see through the heavy steel doorways of the diver's compartment into the lookout room, where there was a single round eye of light.

I climbed up the ladder of the conning-tower and looked out through one of the glass ports. My eyes were just even with the surface of the water. A wave came driving and foaming entirely over the top of the vessel, and I could see the curiously beautiful sheen of the bright summit of the water above us. It was a most impressive sight. Mr. Lake told me that in very clear water it was difficult to tell just where the air left off and the water began; but in the muddy bay where we were going down the surface looked like a peculiarly clear, greenish pane of glass moving straight up and down, not forward, as the waves appear to move when looked at from above.

The latest French aircraft gun

Now we were entirely under water. The rippling noises that the waves had made in beating against the upper structure of the boat had ceased. As I looked through the thick glass port, the water was only three inches from my eyes, and I could see thousands of dainty, semi-translucent jellyfish floating about as lightly as thistledown. They gathered in the eddy behind the conning-tower in great numbers, bumping up sociably against one another and darting up and down with each gentle movement of the water. And I realized that we were in the domain of the fishes.

Jim brought the government chart, and Mr. Lake announced that we were heading directly for Sandy Hook and the open ocean. But we had not yet reached the bottom, and John was busily opening valves and letting in more water. I went forward to the little steel cuddy-hole in the extreme prow of the boat, and looked out through the watch-port. The water had grown denser and yellower, and I could not see much beyond the dim outlines of the ship's spar reaching out forward. Jim said that he had often seen fishes come swimming up wonderingly to gaze into the port. They would remain quite motionless until he stirred his head, and then they vanished instantly. Mr. Lake has a remarkable photograph which he took of a visiting fish, and Wilson tells of nurturing a queer flat crab for days in the crevice of one of the view-holes.

At that moment, I felt a faint jolt, and Mr. Lake said that we were on the bottom of the sea.

Here we were running as comfortably along the bottom of Sandy Hook Bay as we would ride in a Broadway car, and with quite as much safety. Wilson, who was of a musical turn, was whistling Down Went McGinty, and Mr. Lake, with his hands on the pilot-wheel, put in an occasional word about his marvellous invention. On the wall opposite there was a row of dials which told automatically every fact about our condition that the most nervous of men could wish to know. One of them shows the pressure of air in the main compartment of the boat, another registers vacuum, and when both are at zero, Mr. Lake knows that the pressure of the air is normal, the same as it is on the surface, and he tries to maintain it in this condition. There are also a cyclometer, not unlike those used on bicycles, to show how far the boat travels on the wheels; a depth gauge, which keeps us accurately informed as to the depth of the boat in the water, and a declension indicator. By the long finger of the declension dial we could tell whether we were going up hill or down. Once while we were out, there was a sudden, sharp shock, the pointer leaped back, and then quivered steady again. Mr. Lake said that we had probably struck a bit of wreckage or an embankment, but the *Argonaut* was running so lightly that she had leaped up jauntily and slid over the obstruction.

We had been keeping our eyes on the depth dial, the most fascinating and interesting of any of the number. It showed that we were going down, down, down, literally down to the sea in a ship. When we had been submerged far more than an hour, and there was thirty feet of yellowish green ocean over our heads, Mr. Lake suddenly ordered the machinery stopped. The clacking noises of the dynamo ceased, and the electric lights blinked out, leaving us at once in almost absolute darkness and silence. Before this, we had found it hard to realize that we were on the bottom of the ocean; now it came upon us suddenly and not without a touch of awe. This absence of sound and light, this unchanging motionlessness and coolness, this absolute negation—that was the bottom of the sea. It lasted only a moment, but in that moment we realized acutely the meaning and joy of sunshine and moving winds, trees, and the world of men.

A minute light twinkled out like a star, and then another and another, until the boat was bright again, and we knew that among the other wonders of this most astonishing of inventions there was storage electricity which would keep the boat illuminated for hours, without so much as a single turn of the dynamo. With the stopping of the engine, the air supply from above had ceased; but Mr. Lake laid his hand on the steel wall above us, where he said there was enough air compressed to last us all for two days, should anything happen. The possibility of "something happening" had been lurking in our minds ever since we started. "What if your engine should break down, so that you couldn't pump the water out of the water compartments?" I asked. "Here we have hand-pumps," said Mr. Lake promptly; "and if those failed, a single touch of this lever would release our iron keel, which weighs 4000 pounds, and up we would go like a rocket."

I questioned further, only to find that every imaginable

contingency, and some that were not at all imaginable to the uninitiated, had been absolutely provided against by the genius of the inventor. And everything from the gasoline engine to the hand-pump was as compact and ingenious as the mechanism of a watch. Moreover, the boat was not crowded; we had plenty of room to move around and to sleep, if we wished, to say nothing of eating. As for eating, John had brought out the kerosene stove and was making coffee, while Jim cut the pumpkin pie. "This isn't Delmonico's," said Jim, "but we're serving a lunch that Delmonico's couldn't serve—a submarine lunch."

By this time the novelty was wearing off and we sat there, at the bottom of the sea, drinking our coffee with as much unconcern as though we were in an up-town restaurant. For the first time since we started, Mr. Lake sat down, and we had an opportunity of talking with him at leisure. He is a stout-shouldered, powerfully built man, in the prime of life—a man of cool common sense, a practical man, who is also an inventor. And he talks frankly and convincingly, and yet modestly, of his accomplishment.

Having finished our lunch, Mr. Lake prepared to show us something about the practical operations of the *Argonaut*. It has been a good deal of a mystery to us how workmen penned up in a submarine boat could expect to recover gold from wrecks in the water outside, or to place torpedoes, or to pick up cables. "We simply open the door, and the diver steps out on the bottom of the sea," Mr. Lake said, quite as if he was conveying the most ordinary information.

At first it seemed incredible, but Mr. Lake showed us the heavy, riveted door in the bottom of the diver's compartment. Then he invited us inside with Wilson, who, besides being an engineer, is also an expert diver. The massive steel doors of the little room were closed and barred, and then Mr. Lake turned a cock and the air rushed in under high pressure. At once our ears began to throb, and it seemed as if the drums would burst inward.

"Keep swallowing," said Wilson, the diver.

As soon as we applied this remedy, the pain was relieved, but the general sensation of increased air pressure, while exhilarating, was still most uncomfortable. The finger on the pressure dial kept creeping up and up, until it showed that the air pressure inside of the compartment was nearly equal to the water pressure without. Then Wilson opened a cock in the door. Instantly the water gushed in, and for a single instant we expected to be drowned there like rats in a trap. "This is really very simple," Mr. Lake was saying calmly. "When the pressure within is the same as that without, no water can enter."

With that, Wilson dropped the iron door, and there was the water and the muddy bottom of the sea within touch of a man's hand. It was all easy enough to understand, and yet it seemed impossible, even as we saw it with our own eyes. Mr. Lake stooped down, and picked up a wooden rod having a sharp hook at the end. This he pulled along the bottom....

We were now rising again to the surface, after being submerged for more than three hours. I climbed into the conning-tower and watched for the first glimpse of the sunlight. There was a sudden fluff of foam, the ragged edge of a wave, and then I saw, not more than a hundred feet away, a smack bound toward New York under full sail. Her rigging was full of men, gazing curiously in our direction, no doubt wondering what strange monster of the sea was coming forth for a breath of air.

CHAPTER 14

The Modern Submarine

Holland and Lake must be considered the fathers of the modern submarine. This claim is not made in a spirit of patriotic boastfulness, though, of course it is true that the latter was an American by birth, and the former by choice, and that, therefore, we, as a nation, have a right to be proud of the accomplishments of these two fellow-citizens of ours. Without wishing to detract anything from the value of the work done by many men in many countries towards the development of the submarine after and contemporaneously with Holland and Lake, it still remains true that the work which these two did formed the foundation on which all others built. To-day, no submarine worthy of the name, no matter where it has been built and no matter where and how it is used, is without some features which are typical of either the Holland or Lake type. In many instances, and this is true especially of submarines of the highest type and the greatest development, the most significant characteristics of the Holland and Lake boats have been combined.

During the years that followed the small beginnings of Holland and Lake, vast and highly efficient organizations have been built up to continue and elaborate their work. Death claimed Mr. Holland shortly after the outbreak of the great war, on August 12, 1914. Mr. Lake in 1917 was still personally connected with and the guiding spirit of the extensive industrial establishments which have been created at Bridgeport, Conn., as a result of his inventions. He, too, surrounded himself with a corps of experts who in co-operation with him have brought the Lake submarines to a point of perfection which at the time of the *Argonaut*'s first trip would have appeared all but impossible.

Roughly speaking, the beginning of the twentieth century may be called the turning point in the history of submarine invention and the beginning of the modern submarine. Although, as we have heard, various governments, especially those of France and the United States, interested themselves in the submarine question and appropriated small sums of money towards its solution previous to 1900, it was only after that year that governmental interest and influence were set to work with determination and purpose on behalf of submarine inventors. Quite naturally this resulted in increased popular interest. Experimental work on and with submarines no longer had to rely exclusively on private capital, frequently inconveniently timid and limited, but could count now on the vast financial resources of all the great nations of the world. This also made available the unlimited intellectual resources of serious scientists in every part of the universe. Mechanical and electrical engineers, naval designers and constructors, active men of finance and business, and quiet thinkers and investigators in laboratories began to interest themselves in the further development of the submarine.

The United States for a number of years after its adoption of the Holland type remained true to its first choice. Between 1900, when the first Holland boat was bought by the United States Government, and 1911 all the United States submarine, boats were of the Holland type. In the latter year, however, it was decided to give the Lake boat a trial and since that time a number of boats of this type have been built. In all essential features both the Holland and Lake boats of later days were very similar to the original boats of these two types. In all the details, however, immense progress was made. Each new boat thus became greatly superior to its predecessors. This was especially true in regard to size and speed and the improvements made in these two respects naturally resulted in a corresponding increase in radius of activity. The passing years also brought a wonderful refinement of all the technical details of the submarine boats. Practically every feature was developed to a remarkable degree. There is, indeed, a great difference between the submarine boats of the early twentieth century which had to rely on their conning-tower for steering, and more recent boats with their wonderful periscopes and gyro compasses. Similar progress was made in the development of the means of propulsion. The engines used for surface travelling became more powerful and efficient. This was also true of the electric motors, batteries, and accumulators employed in the submerged state. The problem of ventilation likewise has been worked out to such an extent that in the most modern submarines most of the inconveniences experienced by the crews of

Modern German aeroplane types

earlier boats have been removed. This perfection of technical details which was thus gradually approached also permitted a very considerable increase in the fighting power of submarine boats. The number of torpedo tubes was increased and it became possible to carry a larger reserve stock of torpedoes. Submarines of to-day furthermore carry guns varying in calibre, attaining in some instances four inches, and when in later years it became evident that one of the most dangerous enemies of the submarine was the aeroplane, some of the boats were equipped even with anti-aircraft guns.

In the United States Navy the submarine has never been popular. Indeed it is by no means certain that in comparison with other navies of the world the United States was not better off in underwater boats in 1911 than she was three years later when the war-cloud broke. The bulk of our naval opinion has always been for the dreadnoughts. A change of political administration at Washington in 1912 gave a temporary setback to naval development, and the submarines, being still a matter of controversy, languished. Few were built and of those few many showed such structural weakness that the reports of their manoeuvres were either suppressed, or issued in terms of such broad generality that the public could by no possibility suspect, what all the Navy knew to be the fact, that the submarine flotilla of the United States was weak to the point of impotence.

Happily we had nearly three years in which to observe the progress of the war before becoming ourselves embroiled in it. During this period our submarine fleet was somewhat increased, and upon our actual entrance upon the struggle a feverish race was begun to put us on an equality with other nations in underwater boats. It would have been too late had any emergency arisen. But Germany had no ships afloat to be attacked by our submarines had we possessed them. Her own warfare upon our merchant shipping could not be met in kind, for submarines cannot fight submarines. We have, therefore, up to the present time, not suffered from the perilous neglect with which we long treated this form of naval weapon.

Indeed the submarine fleet of the United States Navy at the beginning of the war was so inconsiderable that foreign writers on the subject ignored it. In 1900 we had purchased nine of the type of submarines then put out by the Holland Company. One of these, the first in actual service, known as the "Baby" Holland was kept in commission ten years and upon becoming obsolete was honoured by being taken in state to the Naval Academy at Annapolis and there mounted on a pedestal for the admiration of all comers. She was 59 feet long and would make a striking exhibit placed next to one of the new German submersible cruisers which exceed 300 feet and have a displacement of 5000 tons. These first Holland ships which long constituted the entire underwater force of the United States were but trivial affairs compared with the modern vessel. Their displacement was but 122 tons, their engines for surface navigation were of 160 horse-power, gasoline, and for underwater navigation 70 horse-power, electric. They carried but one torpedo tube and two extra torpedoes and had a radius of action of but 300 miles. At that time in fact the naval theory was that submarines were coast defence vessels altogether. After this war they are likely to form part of the first battle line of every navy. Yet these pioneer vessels established their seaworthiness well in 1911, when four of them accompanied by a parent ship to supply them with fresh stocks of fuel and to render assistance in case of need, crossed the Pacific Ocean under their own power to the Philippines. This exploit tended to popularize these craft in the Navy Department, and soon after larger vessels known as the "Viper" class were ordered. One of these was called the *Octopus*, the first submarine to be fitted with twin screws. In many ways she represented a distinct advance in the art of submarine construction. She was in fact the first vessel built with the distinct idea of being a cruising, as well as a harbour defence ship. Her type proved successful in this respect. The *Octopus* further established a record for deep sea submergence in 1907 when she descended to a depth of 205 feet off Boston, returning to the surface in entire safety.

The ability to withstand the pressure of the water at great depths is a vital quality of a successful submarine. One American submarine narrowly escaped destruction because of structural weakness in this respect. She had by accident descended a few feet below the normal depth at which such boats navigate. The water pressure affected the valves which refused to work and the vessel slowly sank deeper and deeper. At a recorded depth of 123 feet the sinking of the vessel became so much more rapid that the crew with frantic endeavours sought at once to stop the leaks and pump out the water which had entered. At that depth there was a pressure of 153½ pounds upon every square inch of the surface of the submarine. This the workers at the one hand pump had to overcome. It was a savage and a desperate struggle but the men finally won and the vessel regained the surface. As a result of this experience every navy prescribed submergence tests for its submarines before putting them into commission. How to make these tests was perplexing at first. A government did not want to send men down in a steel casket to see just how far they could go before it collapsed. But if no observer accompanied the ship it would be impossible to tell at what depth leakage and other signs of weakness became apparent. An Italian naval architect, Major Laurenti, whose submarines are now found in every navy of the world, invented a dock in which these tests can be made up to any desired pressure while the observers inside the submarine are in communication with those without and the pressure can be instantly removed if signs of danger appear. In the United States Navy boats to be accepted must stand a pressure equivalent to that encountered at 200 feet. In the German navy the depth prescribed is 170 feet. Under normal conditions submarines seldom travel at a depth of more than 100 feet although the *F-1* of the United States Navy accomplished the remarkable feat of making a six-hour cruise in San Francisco Bay at a depth of 283 feet. At this depth the skin of the ship has to withstand a pressure of no less than 123 pounds per square inch.

Specific information as to the nature of submarine construction in the United States since the beginning of the war in 1914 is jealously guarded by the Navy Department. In broad general terms the number of ships under construction is revealed to the public, but all information as to the size of individual vessels, their armour or the qualities of novelty with which every one hopes and believes American inventive genius has invested them, are kept secret. The *Navy Year Book of 1916* summarized our submarine strength at that time as follows:

		Displacement Tons
Submarines fit for action	42	15,722
Under construction	33	21,093
Authorized and appropriated for	30	22,590
Total	105	59,405

In addition thirty-seven more had been authorized by Congress without the appropriation of money for them. By this time however these appropriations have been made together with further heavy ones. While figures are refused at the Navy Department, it is declared that while the United States in 1914 was the last of the great powers in respect to submarine strength provided for, it is now well up to the foremost, even to Germany.

Great Britain like the United States continued for many years to build submarines of the Holland type. Naturally all the recent improvements were incorporated in the British boats. Very little, however, is known concerning the details of the more recent additions to the British submarine flotilla because of the secrecy maintained by the British authorities in war time.

At the beginning of the present war, the British navy possessed 82 active submarines of 5 different classes. They were all of the Holland type, but in each class there were incorporated vast improvements over the preceding class. Displacement, size, motive power, speed, radius of action, and armament were gradually increased until the "E" class contained boats possessing the following features: Submerged displacement, 800 tons; length 176 feet; beam 22½ feet; heavy oil engines of 2000 H.-P.; electric engines of 800 H.-P.; surface speed 16 knots; submerged speed 10 knots; cruising range 5000 miles; armament: 4 torpedo tubes, space for 6 torpedoes, and two 3-inch quick-firing, high-angle, disappearing guns; armoured conning-towers and decks; wireless equipment; 3 panoramic periscopes.

At the same time 22 other submarines were said to be in course of construction. Some of these were of the "F" class (Holland type), similar to the "E" class except that every single characteristic had been greatly increased, in many instances even doubled. In addition to the "F" class Holland-type boats, there were also under construction a number of boats of different types designated respectively as "V," "W," and "S" class. The "V" class were of the Lake type, the "W" of the French "Laubeuf" type, and the "S" class of the Italian "F. I. A. T." or Laurenti type; both of the last named were adaptations of the Lake type.

France, which was for many years the prodigal of the nations when it came to submarine building has continued this tendency. In a way this liberal expenditure of money did not pay particularly well. For, although it resulted in the creation of a comparatively large submarine fleet, this fleet contained boats of every kind and description. Quite a number of the boats were little more than experiments and possessed not a great deal of practical value. The manning and efficient handling of a fleet having so little homogeneity naturally was a difficult matter and seriously restricted its fighting efficiency.

At the outbreak of the war France had 92 submarines in active service, belonging to 12 different classes. In addition there had also been built at various times 5 experimental boats which had been named: *Argonaute, Amiral Bourgeoise, Archimède, Mariotte,* and *Charles Brun.* The majority of the boats belonging to the various classes were of the Laubeuf type, an adaptation of the Lake type made for the French navy by M. Laubeuf, a marine engineer. In their various details these boats vary considerably. Their displacement ranges from 67 tons to 1000 tons, their length from 100 feet to 240 feet, their beam from 12 feet to 20 feet, their surface speed from 8½ knots to 17½ knots, their submerged speed from 5 knots to 12 knots, the horse-power of their heavy oil engines from 1300 to 2000 and that of their electric motors from 350 to 900. Some of the boats, however, have steam engines, others gasoline motors, and still others steam turbines. The cruising range of the biggest and newest boats is 4000 miles. Armament varies with size, of course, the latest boats carrying 4 torpedo tubes for eight 18-inch torpedoes and two 14-pdr. quick-firing, high angle, disappearing guns.

Nine more submarines were in course of construction at the outbreak of war, most of which were of the improved "Gustave Zédé" class. During the war French shipyards were chiefly occupied with capital navy ships and it is not thought the submarine strength has been much increased.

Of the great naval powers, Germany was, strangely enough, the last to become interested in the building of a submarine fleet. This, however, was not due to any neglect on the part of the German naval authorities. It is quite evident from the few official records which are available that they watched and studied very carefully the development of the submarine and growth of the various submarine fleets. During the early years of the twentieth century, however, the Germans seemed to think that most of the boats that were being built then had not yet passed through the experimental stage and they also apparently decided that it would be just as well to wait until other nations had spent their money and efforts on these quasi experimental boats. Not until submarines had been built in the United States, England, and France which had proved beyond all doubt that they were practicable vessels of definite accomplishments, did the Germans seriously concern themselves with the creation of a German submarine fleet. When this period had been reached they went ahead with full power, and with the usual German thoroughness they adopted the best points from each of the various types developed by that time. The result of this attitude was a submarine boat built at first exclusively by Krupp and known as the "Germania" type. It was this type which formed the basis of the German submarine which has become known so extensively and disastrously during recent years. In most respects this type is perhaps more similar to the Lake type than to any other, although some features of the Holland type have been incorporated as well.

GERMAN SUBMARINE MINE-LAYER CAPTURED BY THE BRITISH

At the beginning of the war Germany was credited with only thirty submarines. Six more were then rapidly approaching completion and the German naval law passed some time before provided for the building of seventy-two submarines by the end of 1917. It is believed in fact that by that time the Germans had not less than two hundred *Unterseeboots*.

From the very beginning the Germans have designated their submarines by the letter "U" (standing for *Unterseeboot*) followed by numbers. The first boat was built in 1905 and was named *U-1*. It was a comparatively small boat of 236 tons displacement. The motive power on the surface was a heavy-oil engine of 250 H.-P. Under water the boat was driven by electric motors of a little more than 100 H.-P. Submerged the *U-1* was capable of a speed of 7 knots only, which on the surface of the water could be increased to 10. Her radius of action was about 750 miles. Only one torpedo tube had been provided.

From this boat to the modern German submarine was indeed a long step taken in a comparatively short time. Not very much is known regarding modern German submarines, but the latest boats completed before the war were vessels of 900 tons displacement with heavy-oil engines of 2000 H.-P. and electric motors of 900 H.-P., possessing a surface and submerged speed of 18 and 10 knots respectively and a cruising radius of 4000 miles. They had four torpedo tubes for eight torpedoes, two 14-pdr. quick-firing guns, and two 1-pdr. high-angle anti-aircraft guns. Naturally they were also equipped with all the latest improvements, such as wireless apparatus, panoramic periscopes, armoured conning-towers, and decks. Since the outbreak of the war the Germans have built even more powerful submarine boats whose perfections in regard to speed, radius of action and armament became known through their accomplishments. Of these we will hear more in a later chapter.

At just what period of the war the Germans woke up to the vital importance to them of an enormous submarine fleet is not known. It may have been immediately upon the amazing exploit of Captain Weddigen in the North Sea. At any rate the war had not long progressed before the destruction caused by German submarine attacks began to awaken the apprehension of the Allies and neutral nations. Retaliation in kind was impossible. The Germans had neither merchant nor naval ships at sea to be sunk. The rapidity with which the volume of the loss inflicted upon merchant shipping grew indicated an equally rapid increase in the size of the German underwater fleet. Neutrals were enraged by the extension by the Germans of the areas of sea in which they claimed the right to sink neutral ships, and their growing disregard for the restraining principles of international law. How greatly they developed the submarine idea was shown by their construction in 1916 of vessels with a displacement of 2400 tons; a length of 279 feet, and a beam of 26 feet; a surface speed of 22 knots, cruising radius of 6500 miles, mounting 4 to 8 guns and carrying a crew of from 40 to 60. But it was reported that two vessels designed primarily for surface cruising, but nevertheless submersible at will, had been laid down of 5000 tons, a length of 414 feet, and a radius of 18,000 to 20,000 miles. These "submersible cruisers" as they were called, mounted 6 to 8 guns, 30 torpedo tubes, and carried 90 torpedoes. What part vessels of this type shall play in war is still to be determined.

Of the smaller naval powers, Italy comparatively early had become interested in the building of submarines. Most of her boats are of the Laurenti type—which is a very close adaptation of the Lake type. Russia and Japan, especially the latter, built up fairly efficient underwater fleets. The lesser countries, like Austria, Holland, Sweden, Denmark, Norway, and Spain have concerned themselves seriously with the creation of submarine fleets. The submarine boats of all of these countries in most instances were either of the Lake or Holland type though frequently they were built from plans of English, French or German adaptations rather than in accordance with the original American plans.

The exact number of submarines possessed now by the various navies of the world is a matter of rather indefinite knowledge. Great secrecy has been maintained by every country in this respect. From a variety of sources, however, it has been possible to compile the following list which at least gives an approximate idea of the respective strength of the various submarine fleets at the beginning of the war. The numbers assigned to each country are only approximate, however, and include both boats then in existence or ordered built: United States 57; Great Britain 104; France 92; Germany 36; Italy 28; Russia 40; Japan 15; Austria 12; Holland 13; Denmark 15; Sweden 13; Norway 4; Greece 2; Turkey 2; Brazil 3; Peru 2.

Having traced the development of the submarine from its earliest beginnings to recent times we are naturally now confronted with the question "What are the principal requirements and characteristics of the modern submarine?"

The submarine boat of to-day, in order to do its work promptly and efficiently, must first of all possess seaworthiness. This means that no matter whether the sea is quiet or rough the submarine must be able to execute its operations with a fair degree of accuracy and promptness and must also be capable of making continuous headway. Surface and underwater navigation must be possible with equal facility and it is necessary that a state of submergence can be reached without loss of time and without any degree of danger to the boat's safety. At all times, travelling above water or below, the submarine must possess mechanical means which will make it possible to control its evolutions under all conditions. Furthermore, the ability of the submarine to find and to observe objects in its vicinity must not be greatly reduced when it is in a submerged position. In the latter it also becomes of extreme importance that the provisions for ventilation are such that the crew of the submarine should lose as little as possible in its efficiency and comfort. A fair amount of speed both on and below the surface of the water is essential and the maintenance of the speed for a fairly long period of time must be assured.

In regard to their general outward appearance, submarines of various types to-day vary comparatively little. In many respects they resemble closely in shape, torpedo boats—the earlier submarines particularly. In size, of course, they differ in accordance with the purposes for which they have been designed. As compared with earlier submarines the most notable difference is that modern submarines possess more of a superstructure. Almost all of them are built now with double hulls. The space between the outer and the inner hull is utilized primarily for ballast tanks by means of which submergence is accomplished and stability maintained and regulated. Some of these tanks, however, are not used to carry water ballast, but serve as reservoirs for the fuel needed by the engines. The stability of the submarine and the facility with which it can submerge also depend greatly on the distribution of weight of its various parts. This problem has been worked out in such a way that to-day there is little room for improvement. Its details, however, are of too technical a nature to permit discussion in this place.

Hydroplanes both fore and aft are now generally used to assist in regulating and controlling stability in the submerged state. The motive power of the modern submarine is invariably of a two-fold type. For travelling on the surface internal combustion engines are used. The gasoline engine of former years has been displaced by Diesel motors or adaptations of them. Although these represent a wonderful advance over the engines used in the past there is still a great deal of room for improvement. The opinions of engineers in this respect vary greatly, American opinion being generally unfavourable to the Diesel type, and whether the final solution of this problem will lie in the direction of a more highly developed motor of Diesel type, of an improved gasoline engine, or of some other engine not yet developed, only the future can tell. Simplicity of construction and reliability of operation are the two essential features which must be possessed by every part of the power plant of a submarine. For underwater travel electric motors and storage batteries are employed exclusively. These vary, of course, in detail. In principle, however, they are very much alike. Although this combination of electric and oil power is largely responsible for having made the submarine what it is to-day, it is far from perfect. Mechanical complications of many kinds and difficulties of varying degrees result from it. Up to comparatively recently these were considered insurmountable obstacles. But engineers all over the world are giving their most serious attention to the problem of devising a way to remove these obstacles and continuous progress is made by them.

As an immediate result of the development of motive power in the submarine its speed both on and below the surface of the water as well as its radius of action has been materially increased. To-day submarines travel on the water with a speed which even a few years ago would have been thought quite respectable for the most powerful battleships or the swiftest passenger liners. And even under water, submarines attain a velocity which is far superior to that of which earlier submarines were capable on the surface of the water. How immensely extended the radius of action of the submarine has become in recent years, has impressed itself on the world especially in the last few years. Both English and French submarines have travelled without making any stops from their home ports to the Dardanelles and back again. And used to, and satiated as we are with mechanical wonders of all kinds the whole world was amazed when in 1916 German submarines made successful trips from their home ports to ports in the United States and returned with equal success. This meant a minimum radius of action of 3500 miles. In the case of the German U-boat which in 1916 appeared at Newport for a few hours, then attacked and sank some merchantmen off the United States coast and later was reported as having arrived safely in a German port, it has never been established whether the boat renewed its supplies of food and fuel on the way or carried enough to make the trip of some 7000 miles.

One other important feature without which submarines would have found it impossible to score such accomplishments is the periscope. In the beginning periscopes were rather crude appliances. They were very weak and sprung leaks frequently. Moisture, formed by condensation, made them practically useless. In certain positions the image of the object picked up by the periscope became inverted. Their radius of vision was limited, and in every way they proved unreliable and unsatisfactory. But, just as almost every feature of submarine construction was gradually developed and most every technical obstacle overcome, experts gradually concentrated their efforts on the improvement of periscopes. Modern periscopes are complicated optical instruments which have been developed to a very high point of efficiency. A combination of prisms and lenses makes it possible now to see true images clearly. Appliances have been developed to make the rotation of the periscope safe, prompt, and easy so that the horizon can be swept readily in every direction. Magnification can be established at will by special devices easily connected or disconnected with the regular instrument. The range of vision of the modern periscope is as remarkable as its other characteristics. It differs, of course, in proportion to the height to which the periscope is elevated above the surface of the water. In clear weather a submarine, having elevated its periscope to a height of 20 feet can pick up a large battleship at as great a distance as 6 miles, while observers on the latter, even if equipped with the most powerful optical instruments, are absolutely unable to detect the submarine. This great distance is reduced to about 4000 yards if the periscope is only 3 feet above the surface of the water and to about 2200 yards if the elevation of the periscope is 1 foot. But even the highly developed periscope of to-day, usually called "panoramic periscope," has its limitations. The strain on the observer's eyes is very severe and can be borne only for short periods. In dirty weather the objectives become cloudy and the images are rendered obscure and indefinite, although this trouble has been corrected, at least in part, by forcing a strong blast through the rim surrounding the observation glass. At night, of course, the periscope is practically useless. Formerly a shot which cut off the periscope near the water's edge might sink the boat. This has been guarded against by cutting off the tube with a heavy plate of transparent glass which does not obstruct vision but shuts off the entrance of water.

Important as the periscope is both as a means of observing the surroundings of the submarine and as a guide in steering it, it is not

The exterior of first German submarine

the only means of accomplishing the latter purpose. To-day every submarine possesses the most reliable type of compass available. At night when the periscope is practically useless or in very rough weather, or in case the periscope has been damaged or destroyed, steering is done exclusively by means of the compass. The latest type in use now on submarines is called the gyroscope compass which is a highly efficient and reliable instrument.

In the matter of ventilation the modern submarine also has reached a high state of perfection. The fresh air supply is provided and regulated in such a manner that most of the discomforts suffered by submarine crews in times past have been eliminated. The grave danger which formerly existed as a result of the poisonous fumes, emanating from the storage batteries and accumulators, has been reduced to a minimum. In every respect, except that of space, conditions of life in a submarine have been brought to a point where they can be favourably compared with those of boats navigated on the surface of the water. Of course, even at the best, living quarters in a submarine will always be cramped. However, it is so important that submarine crews should be continuously kept on a high plane of efficiency that they are supplied with every conceivable comfort permitted by the natural limitations of submarine construction.

Submarine boats so far have been used almost exclusively as instruments of warfare. One of their most important features, therefore, naturally is their armament. We have already heard something about the use of torpedoes by submarines. The early submarines had as a rule only one torpedo tube and were incapable of carrying more than two or three torpedoes. Gradually, however, both the number of torpedo tubes and of torpedoes was increased. The latest types have as many as eight or ten tubes and carry enough torpedoes to permit them to stay away from their base for several weeks. In recent years submarines have also been armed with guns. Naturally these have to be of light weight and small calibre. They are usually mounted so that they can be used at a high angle. This is done in order to make it possible for submarines to defend themselves against attacks from airships. The mountings of these guns are constructed in such a way that the guns themselves disappear immediately after discharge and are not visible while not in use. Though mounted on deck they are aimed and fired from below. As part of the armament of the submarine we must also consider the additional protection which they receive from having certain essential parts protected by armour plate.

The interior of first German submarine, showing appliances for man-power

All these features have increased the safety of submarine navigation to a great extent. In spite of the popular impression that submarine navigation entailed a greater number of danger factors than navigation on the surface of the water, this is not altogether so. If we stop to consider this subject we can readily see why rather the opposite should be true. Navigation under the surface of the water greatly reduces the possibility of collision and also the dangers arising from rough weather. For the results of the latter are felt to a much lesser degree below than on the surface of the water. Many other factors are responsible for the comparatively high degree of safety inherent in submarines. Up to the outbreak of the present war only about two hundred and fifty lives had been lost as a result to accidents to modern submarines. Considering that up to 1910 a great deal of submarine navigation was more or less experimental this is a record which can bear favourable comparison with similar records established by overwater navigation or by navigation in the air.

To the average man the thought of imprisonment in a steel tube beneath the surface of the sea, and being suddenly deprived of all means of bringing it up to air and light is a terrifying and nerve shattering thing. It is probably the first consideration which suggests itself to one asked to make a submarine trip. Always the newspaper headlines dealing with a submarine disaster speak of those lost as "drowned like rats in a trap." Men will admit that the progress of invention has greatly lessened the danger of accident to submarines, but nevertheless sturdily insist that when the accident does happen the men inside have no chance of escape.

As a matter of fact many devices have been applied to the modern submarine to meet exactly this contingency. Perhaps nothing is more effective than the so-called telephone buoy installed in our Navy and in some of those of Europe. This is a buoy lightly attached to the outer surface of the boat, containing a telephone transmitter and receiver connected by wire with a telephone within. In the event of an accident this buoy is released and rises at once to the surface. A flag attached attracts the attention of any craft that may be in the neighbourhood and makes immediate communication with those below possible. Arrangements can then be made for raising the boat or towing her to some point at which salvage is possible. An instance of the value of this device was given by the disaster to the German submarine *U-3* which was sunk at Kiel in 1910. Through the telephone the imprisoned crew notified those at the other end that they had oxygen enough for forty-eight hours but that the work of rescue must be completed in that time. A powerful floating derrick grappled the sunken submarine and lifted its bow above water. Twenty-seven of the imprisoned crew crept out through the torpedo tubes. The captain and two lieutenants conceived it their duty to stay with the ship until she was actually saved. In the course of the operations one of the ventilators was broken, the water rushed in and all three were drowned.

In some of the Holland ships of late construction there is an ingenious, indeed an almost incredible device by which the ship takes charge of herself if the operators or crew are incapacitated. It has happened that the shock of a collision has so stunned the men cooped up in the narrow quarters of a submarine that they are for quite an appreciable time unable to attend to their duties. Such a collision would naturally cause the boat to leak and to sink. In these newer Holland ships an automatic device causes the ship, when she has sunk to a certain depth, registered of course by automatic machinery, to start certain apparatus which empties the ballast tanks and starts the pumps which will empty the interior of the ship if it has become flooded. The result is that after a few minutes of this automatic work, whether the crew has sufficiently recovered to take part in it or not, the boat will rise to the surface.

This extraordinary invention is curiously reminiscent of the fact chronicled in earlier chapters of this book that the most modern aeroplanes are so built that should the aviator become insensible or incapacitated for his work, if he will but drop the controls, the machine will adjust itself and make its own landing in safety. Unaided the aeroplane drops lightly to earth; unaided the submarine rises buoyantly to the air.

In recent years there have been developed special ships for the salvage of damaged or sunk submarines. At the same time the navies of the world have also produced special submarine tenders or mother ships. The purpose of these is to supply a base which can keep on the move with the same degree of facility which the submarine itself possesses. These tenders are equipped with air compressors by means of which the air tanks of submarines can be refilled. Electric generators make it possible to replenish the submarine storage batteries. Mechanical equipment permits the execution of repairs to the submarine's machinery and equipment. Extra fuel, substitute parts for the machinery, spare torpedoes are carried by these tenders. The most modern of them are even supplied with dry dock facilities, powerful cranes, and sufficiently strong armament to repel attacks from boats of the type most frequently encountered by submarines.

There are, of course, many other special appliances which make up the sum total of a modern submarine's equipment. Electricity is used for illuminating all parts of the boat. Heat is supplied in the same manner; this is a very essential feature because the temperature of a submarine, after a certain period of submergence, becomes uncomfortably low. Electricity is also used for cooking purposes.

Every submarine boat built to-day is equipped with wireless apparatus. Naturally it is only of limited range varying from one hundred and twenty to one hundred and eighty miles, but even at that it is possible for a submarine to send messages to its base or some other given point from a considerable distance by relay. If the submarine is running on the surface of the water the usual means of naval communication-flag signals, wig-wagging or the semaphore, can be employed. The submarine bell is another means for signalling. It is really a wireless telephone, operating through the water instead of the air. Up to the present, however, it has not been sufficiently developed to permit its use for any great distance. It is so constructed that it can also be used as a sound detector.

Some submarines, besides being equipped with torpedo tubes, carry other tubes for laying mines. In most instances this is only a secondary function of the submarine. There are, however, special mine-laying submarines. Others, especially of the Lake type, have diving compartments which permit the employment of divers for the purpose of planting or taking up mines. Disappearing anchors, operated by electricity from within the boat, are carried. They are used for steadying the boat if it is desired to keep it for any length of time on the bottom of the sea in a current.

From this necessarily brief description it can be seen readily that the modern submarine boat is a highly developed, but very complicated mechanism. Naturally it requires a highly trained, extremely efficient crew. The commanding officers must be men of strong personality, keen intellect, high mechanical efficiency, and quick judgment. The gradual increase in size has brought a corresponding increase in the number of a submarine's crew. A decade ago from 8 to 10 officers and men were sufficient but to-day we hear of submarine crews that number anywhere from 25 to 40.

In spite of the marvellous advances which have been made in the construction, equipment, and handling of the submarine during the last ten years, perfection in many directions is still a long way off. How soon it will be reached, if ever, and by what means, are, of course, questions which only the future can answer.

CHAPTER 15

Aboard a Submarine

Submarines have been compared to all kinds of things, from a fish to a cigar. Life on them has been described in terms of the highest elation as well as of the deepest depression. Their operation and navigation, according to some claims, require a veritable combination of mechanical, electrical, and naval genius—not only on the part of the officers, but even on that of the simplest oiler—while others make it appear as if a submarine was at least as simple to handle as a small motor boat. The truth concerning all these matters lies somewhere between these various extremes.

It is quite true that except on the very latest "submerged cruisers" built by the Germans, the space for the men operating a submarine is painfully straitened. They must hold to their positions almost like a row of peas in a pod. From this results the gravest strain upon the nerves so that it has been found in Germany that after a cruise a period of rest of equal duration is needed to restore the men to their normal condition. Before assignment to submarine duty, too, a special course of training is requisite. Submarine crews are not created in a day.

What the interior of the new German submarines with a length of 280 feet, and a beam of 26 feet may be, no man of the Anglo-Saxon race may know or tell. The few who have descended into those mysterious depths will have no chance to tell of them until the war is over. Nor is it possible during wartimes to secure descriptions even of our own underwater boats. But the interior of the typical submarine may be imagined as in size and shape something like an unusually long street car. Along the sides, where seats would normally be, are packed wheels, cylinders, motors, pumps, machinery of all imaginable kinds and some of it utterly unimaginable to the lay observer. The whole interior is painted white and bathed in electric light. The casual visitor from "above seas" is dazed by the array of machinery and shrinks as he walks the narrow aisle lest he become entangled in it.

Running on the surface the submarine chamber is filled with a roar and clatter like a boiler shop in full operation. The Diesel engines are compact and powerful, but the racket they make more nearly corresponds to their power than to their size. On the surface too the boat rolls and pitches and the stranger passenger, unequipped with sea legs grabs for support as the subway rider reaches for a strap on the curves. But let the order come to submerge. The Diesels are stopped. The electric motors take up the task, spinning noiselessly in their jackets. In a moment or two all rolling ceases. One can hardly tell whether the ship is moving at all—it might for all its motion tells be resting quietly on the bottom. If you could disabuse your mind for a moment of the recollection that you were in a great steel cigar heavy laden with explosives, and deep under the surface of the sea you would find the experience no more exciting than a trip through the Pennsylvania tubes. But there is something uncanny about the silence.

Go forward to the conical compartment at the very bow. There you will find the torpedo chamber for the submarine, like the cigar to which it is so often compared, carries its fire at its front tip. The most common type of boat will have two or four torpedo tubes in this chamber. The more modern ones will have a second torpedo chamber astern with the same number of tubes and carry other torpedoes on deck which by an ingenious device can be launched from their outside cradles by mechanism within the boat. In the torpedo chamber are twice as many spare torpedoes as there are tubes, made fast along the sides. Here too the anchor winch stands with the cable attached to the anchor outside the boat and an automatic knife which cuts the cable should the anchor be fouled.

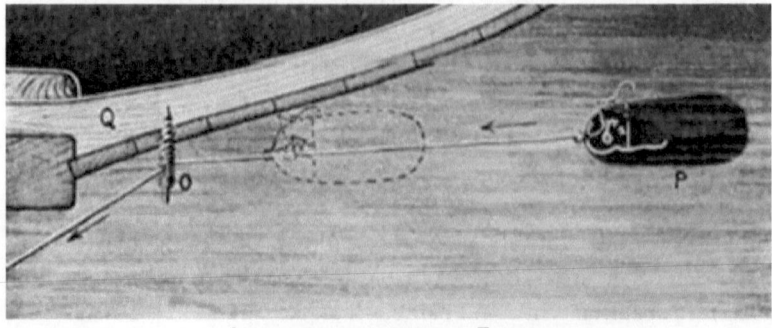

A TORPEDO DESIGNED BY FULTON

Immediately aft of the torpedo chamber, cut off by a water-tight partition, is the battery compartment. It gets its name because of the fact, that beneath the deck which is full of traps readily raised are the electric storage batteries of anywhere from 60 to 260 cells according to the size of the boat. This room is commonly used as the loafing place for the crew, being regarded as very spacious and empty. In it are nothing but the electric stove, the kitchen sink, the various lockers for food and all the housekeeping apparatus of the submarine. Mighty trim and compact they all are. The builder of twentieth century flats with his kitchenettes and his in-door beds might learn a good deal from a study of the smaller type of submarine. Next aft come the officers' staterooms, rather smaller than prison cells, each holding a bunk, a bureau, and a desk. Each holds also a good deal of moisture, for the greatest discomfort in submarine life comes from the fact that everything is dripping with the water resulting from the constant condensation of the air within.

The great compartment amidships given over to machinery is a place to test the nerves. The aisle down the centre is scarcely two feet wide and on each side are whirling wheels, engines, and electric motors. Only the photographs can give a clear idea of the crowded appearance of this compartment. It contains steering wheels, the gyroscopic compass, huge valves, dials showing depth of submergence, Kingston levers, motor controllers, all polished and shining, each doing its work and each easily thrown out of gear by an ignorant touch.

The author once spending the night on a United States man-of-war was shown by the captain to his own cabin, that officer occupying the admiral's cabin for the time. At the head of the bunk were two small electric push buttons absolutely identical in appearance and about two inches apart. "Push this button," said the captain genially, "if you want the Jap boy to bring you shaving water or anything else. But be sure to push the right one. If you push the other you will call the entire crew to quarters at whatever hour of night the bell may ring."

The possibility of mistaking the button rested heavily on the writer's nerves all night. A somewhat similar feeling comes over one who walks the narrow path down the centre of the machinery compartment of a submarine. He seems hedged about by mysterious apparatus a touch of which, or even an accidental jostle may release powerful and even murderous forces.

While the submarine is under way, submerged, the operator at every piece of individual machinery stands at its side ready for ac-

The method of attack by Nautilus

tion. Here are the gunner's mates at the diving rudder. They watch steadily a big gauge on which a needle which shows how deep the boat is sinking. When the required depth is reached swift turns of two big brass wheels set the horizontal rudders that check the descent and keep the boat on an even keel. Other men stand at the levers of the Kingston valves which, when open, flood the ballast tanks with water and secure the submergence of the boat. Most of the underwater boats to-day sink rapidly on an even keel. The old method of depressing the nose of the boat so as to make a literal dive has been abandoned, partly because of the inconvenience it caused to the men within who suddenly found the floor on which they were standing tilted at a sharp angle, and partly because the diving position proved to be a dangerous one for the boat.

In the early days of the submarines the quarters for the men were almost intolerable. The sleeping accommodations were cramped and there was no place for the men off duty to lounge and relax from the strain of constant attention to duty. Man cannot keep his body in a certain fixed position even though it be not rigid, for many hours. This is shown as well at the base ball grounds at the end of the sixth inning when "all stretch" as it was in the old time underwater boats. The crews now have space in which to loaf and even the strain of long silent watches under water is relieved by the use of talking machines and musical instruments. The efficiency of the boat of course is only that of her crew, and since more care and more scientific thought has been given to the comfort of the men, to the purity of the air they breathe, and even to their amusements, the effect upon the work done by the craft has been apparent. Ten years ago hot meals were unthought of on a submarine; now the electric cooker provides for quite an elaborate bill of fare. But ten years ago the submarine was only expected to cruise for a few hours off the harbour's mouth carrying a crew of twenty men or less. Now it stays at sea sometimes for as long as three months. Its crews number often as many as fifty and the day is in sight when accommodations will have to be made for the housing of at least eighty men in such comparative comfort that they can stand a six months' voyage without loss of morale or decrease in physical vigour.

It is, of course, very rare that a civilian has the chance to be present on a submarine when the latter is making either a real or a feigned attack. Fred B. Pitney, a correspondent of the *New York Tribune*, was fortunate enough to have this experience, fortunate especially because it was all a game arranged for his special benefit by a French admiral. He writes of this interesting experience in the *Tribune* of Sunday, May 27, 1917, and at the same time gives a vivid description of a French submarine.

It appears that Mr. Pitney was on a small vessel put at his disposal by the French Ministry of Marine to view the defences of a French naval base. This boat was attacked by what seemed to be an enemy submarine, but later turned out to be a French one which was giving this special performance for Mr. Pitney's information. We read:

> Our officers were experts at watching for submarines, and though the little white wave made by the periscope disappeared, they caught the white wake of the torpedo coming toward the port quarter and sheered off to escape it. The torpedo passed harmlessly by our stern, but the adventure was not ended, for hardly a minute later we heard a shot from off the starboard quarter and, turning in that direction, saw that the submarine had come to the surface and was busily firing at us to bring us to.
>
> We stopped without any foolish waste of time in argument. I asked if a boat would be sent to us, or if we would have to get out our boat.
>
> "They carry a small folding boat," said the officer to whom I had been talking, "but we will have to send our boat."
>
> While we were getting our boat over the side, the submarine moved closer in, keeping her gun bearing on us all the time, most uncomfortably. The gun stood uncovered on the deck, just abaft the turret. It was thickly coated with grease to protect it when the vessel submerged. It is only the very latest type of submarines that have disappearing guns which go under cover when the vessel submerges and are fired from within the ship, which makes all the more surprising the speed with which a submarine can come to the surface, the men get out on deck, fire the gun, get in again and the vessel once more submerges.
>
> I was in the first boatload that went over to the submarine. From a distance it looked like nothing so much as a rather long piece of 4×8 floating on the water, with another block set on top of it and a length of lath nailed on the block. It lost none of these characteristics as we neared it. It only gained a couple of ropes along the sides of the 4×8, while men kept coming mysteriously out of the block until a round dozen

THE CAPTURE OF A U-BOAT

was waiting to receive us. The really surprising thing was that the men turned out to be perfectly good French sailors, with a most exceedingly polite French lieutenant to help us aboard the little craft....

The vessel we were in was a 500-ton cruising submarine. It had just come from eight months' guarding the Channel, and showed all the battering of eight months of a very rough and stormy career with no time for a lie-up for repairs. It was interesting to see the commander hand the depth gauge a wallop to start it working and find out if the centre of the boat was really nine feet higher than either end. We were fifty-four feet under water and diving when the commander performed that little experiment and we continued to dive while the gauge spun around and finally stopped at a place which indicated approximately that our back was not broken. I suppose that was one of the things my friend the lieutenant referred to when he said life on a submarine was such a sporting proposition.

We boarded the submarine over the tail end and balanced our way up the long narrow block, like walking a tight rope, to the turret, where we descended through a hole like the opening into a gas main into a small round compartment about six feet in diameter exactly in the midships section, which was the largest compartment in the ship. Running each way from it the length of the vessel were long corridors, some two feet wide. On each side of the corridors were rows of tiny compartments, which were the living and working rooms of the ship. Naturally, most of the space was given up to the working rooms.

The officers' quarters consisted of four tiny compartments, two on each side of the after corridor. The first two were the mess room and chart room, and the second pair were the cabins of the commander—a lieutenant—and his second in command, an ensign. Behind them was an electric kitchen, and next came the engines, first two sets of Diesel engines, one on each side of the corridor, each of four hundred horse-power. These were for running on the surface. Then came four bunks for the quartermasters and last the electric motors for running under the surface. The motors were run from storage batteries and were half the power of the Diesel engines. The quarters of the crew were along the sides of the forward corridor. The floors of the corridor were an unbroken series of trap doors, covering the storage tanks for drinking water, food, and the ship's supplies. The torpedo tubes were forward of the men's quarters. Ten torpedoes were carried. The ammunition for the deck gun was stored immediately beneath the gun, which was mounted between the turret and the first hatch, abaft the turret. Besides the turret there were three hatches in the deck, one forward and two aft.

There were thirty-four men in the crew. The men are counted every two hours, as there is great danger of men being lost overboard when running on the surface, and in bad weather they are sometimes counted as often as every half hour.

The turret was divided in two sections. In the after part was the main hatch and behind it a stationary periscope, standing about thirty inches above the surface of the water when the deck was submerged and only the periscope showing. There was no opening in the forward section of the turret, but the fighting periscope, which could be drawn down into the interior or pushed up to ten feet above the surface when the vessel was completely submerged, extended through the top.

For two hours, turn and turn about, the commander and his second stand watch on the iron grips in the turret, one eye on the periscope, the other on the compass. And this goes on for weeks on end. It is only when they lie for a few hours fifty to seventy-five feet below the surface that they can get some rest. And even then there is no real rest, for one or the other of them must be constantly on duty, testing pipes and gauges, air pressure, water pressure, and a thousand other things.

When we dropped through the hatch into the interior of the submarine and the cover was clamped down over our heads the commander at once ordered me back into the turret.

"Hurry, if you want to see her dive," he said.

I climbed into the after section of the turret and fastened my eye to the periscope. Around the top of the turret was a circle of bulls' eyes and I was conscious of the water dashing against them while the spray washed over the glass of the periscope. The little vessel rolled very slightly on the surface, though there was quite a bit of sea running. I watched the horizon through the periscope and watched for the dive, expecting a distinct sensation, but the first thing I noticed was that even the slight roll had ceased and I was surprised to see that the bulls' eyes were completely under water. The next thing there was no more horizon. The periscope also was covered and we were completely beneath the surface.

"Did it make you sick?" the commander asked, when I climbed down from the turret, and when I told him "no" he was surprised, for he said most men were made sick by their first dive.

The thing most astonishing to me about that experience was how a submerged submarine can thread its way through a mine field. For though the water is luminous and translucent one can hardly make out the black hull of the boat under the turret and a mine would have to be on top of you before you could see it. The men who watch for mines must have a sense for them as well as particularly powerful sight.

We continued to dive until we were sixty-eight feet below the surface, too deep to strike any mine, and there we ran tranquilly on our electric engines, while the commander navigated the vessel and the second in command opened champagne in the two by four mess room. After half an hour of underwater work we came near enough the surface for our fighting periscope to stick twenty inches out of the water and searched the lonely horizon for a ship to attack.

It was not long before we sighted a mine trawler, steaming for the harbour, and speeded up to overtake her.

"Pikers!" said our commander, as we circled twice around the trawler; "they can't find us."

Five men on the trawler were scanning the sea with glasses looking for submarines. We could follow all their motions, could tell when they thought they had found us and see their disappointment at their mistakes, but though we were never more than five hundred yards from them, I did not think they were pikers because they did not find us. I had tried that hunt for the tiny wave of a periscope.

"No use wasting a torpedo on those fellows," said our commander. "We will use the gun on them."

"How far away can you use a torpedo?" I asked.

"Two hundred yards is the best distance," he said. "Never more than five hundred. A torpedo is pure guesswork at more than five hundred yards."

We crossed the bow of the trawler, circled around to her

starboard quarter and came to the surface, fired nine shots and submerged again in forty-five seconds.

The prey secured, we ran submerged through the mine field and past the net barrier to come to the surface well within the harbour and proceed peacefully to our mooring under the shelter of the guns of the land forts.

Life and work on a German submarine is known to us, of course, only from descriptions in German publications. One of these appeared, previous to our entry in the war, in various journals and was translated and republished by the *New York Evening Post*. It reads partly as follows:

"*U-47* will take provisions and clear for sea. Extreme economical radius."

A first lieutenant, with acting rank of commander, takes the order in the grey dawn of a February day. The hulk of an old corvette with the Iron Cross of 1870 on her stubby foremast is his quarters in port, and on the corvette's deck he is presently saluted by his first engineer and the officer of the watch. On the pier the crew of *U-47* await him. At their feet the narrow grey submarine lies alongside, straining a little at her cables.

"Well, we've our orders at last," begins the commander, addressing his crew of thirty, and the crew grin. For this is *U-47*'s first experience of active service. She has done nothing save trial trips hitherto, and has just been overhauled for her first fighting cruise. Her commander snaps out a number of orders. Provisions are to be taken in "up to the neck," fresh water is to be put aboard, and engine-room supplies to be supplemented.

A mere plank is the gangway to the little vessel. As the commander, followed by his officers, comes aboard, a sailor hands to each a ball of cotton-waste, the sign and symbol of a submarine officer, which never leaves his hand. For the steel walls of his craft, the doors, and the companion-ladder all sweat oil, and at every touch the hands must be wiped dry. The doorways are narrow round holes. Through one of the holes aft the commander descends by a breakneck iron ladder into the black hole lit by electric glow-lamps. The air is heavy with the smell of oil, and to the unaccustomed longshoreman it is almost choking, though the hatches are off. The submarine man breathes this air as if it were the purest ozone. Here in the engine-room aft men must live and strain every nerve even if for days at a time every crack whereby the fresh air could get in is hermetically sealed. On their tense watchfulness thirty lives depend.

Here, too, are slung some hammocks, and in them one watch tries, and, what is more, succeeds in sleeping, though the men moving about bump them with head and elbows at every turn, and the low and narrow vault is full of the hum and purr of machinery. In length the vault is about ten feet, but if a man of normal stature stands in the middle and raises his arms to about half shoulder height his hands will touch the cold, moist steel walls on either side. A network of wires runs overhead, and there is a juggler's outfit of handles, levers, and instruments. The commander inspects everything minutely, then creeps through a hole into the central control station, where the chief engineer is at his post. With just about enough assistance to run a fairly simple machine ashore the chief engineer of a submarine is expected to control, correct, and, if necessary, repair at sea an infinitely complex machinery which must not break down for an instant if thirty men are to return alive to the hulk.

Forward is another narrow steel vault serving at once as engine-room and crew's quarters. Next to it is a place like a cupboard, where the cook has just room to stand in front of his doll's house galley-stove. It is electrically heated, that the already oppressive air may not be further vitiated by smoke or fumes. A German submarine in any case smells perpetually of coffee and cabbage. Two little cabins of the size of a decent clothes-chest take the deck and engine-room officers, four of them. Another box cabin is reserved for the commander—when he has time to occupy it.

At daybreak the commander comes on deck in coat and trousers of black leather lined with wool, a protection against oil, cold, and sea-water. The crew at their stations await the command to cast off.

"Machines clear," calls a voice from the control-station and "Clear ship," snaps the order from the bridge. Then "Cast-off!" The cables slap on to the landing-stage, the engines begin to purr, and *U-47* slides away into open water.

A few cable-lengths away another submarine appears homeward bound. She is the *U-20* returning from a long cruise in which she succeeded in sinking a ship bound with a cargo of frozen mutton for England.

"Good luck, old sheep-butcher," sings the commander of *U-47* as the sister-ship passes within hail.

The seas are heavier now, and *U-47* rolls unpleasantly as she makes the light-ship and answers the last salute from a friendly hand. The two officers on the bridge turn once to look at the light-ship already astern, then their eyes look seaward. It is rough, stormy weather. If the egg-shell goes ahead two or three days without a stop, the officers in charge will get no sleep for just that long. If it gets any rougher they will be tied to the bridge-rails to avoid being swept overboard. If they are hungry, plates of soup will be brought to them on the bridge, and the North Sea will attend to its salting for them.

Frequently this "meal" is interrupted by some announcement from the watch, such as: "Smoke on the horizon off the port bow." Then—so we are told:

The commander drops his plate, shouts a short, crisp command, and an electric alarm whirs inside the egg-shell. The ship buzzes like a hive. Then water begins to gurgle into the ballast-tanks, and *U-47* sinks until only her periscope shows.

"The steamship is a Dutchman, sir," calls the watch officer. The commander inspects her with the aid of a periscope. She has no wireless and is bound for the Continent. So he can come up and is glad, because moving under the water consumes electricity, and the usefulness of a submarine is measured by her electric power.

After fifty-four hours of waking nerve tension, sleep becomes a necessity. So the ballast-tanks are filled and the nutshell sinks to the sandy bottom. This is the time for sleep aboard a submarine, because a sleeping man consumes less of the precious oxygen than one awake and busy. So a submarine man has three principal lessons to learn—to keep every faculty at tension when he is awake, to keep stern silence when he is ashore (there is a warning against talkativeness in all the German railway-carriages now), and to sleep instantly when he gets a legitimate opportunity. His sleep and the economy of oxygen may save the ship. However, the commander allows half an hour's grace for music. There is a gramophone, of course, and the "ship's band" performs on all manner of instruments. At worst, a comb with a bit of tissue paper is pressed into service.

Another American who suffered an enforced voyage on an *unterseeboot* made public later some of his experiences. His captor's craft was a good sized one—about 250 feet long, with a crew of 35 men and mounting two 4½ inch guns. She could make 18 knots on the surface and 11 submerged and had a radius of 3200 miles of action. Her accommodations were not uncomfortable. Each officer had a separate cabin while the crew were bunked along either side of a narrow passage. The ventilation was excellent, and her officers declared that they could stand twenty-four hours continuous submergence without discomfort, after that for six hours it was uncomfortable, and thereafter intolerable because of the exudation of moisture—or sweating—from every part. At such times all below have to wear leather suits. The food was varied and cooked on an electric stove. The original stores included preserved pork and beef, vegetables, tinned soups, fruits, raisins, biscuits, butter, marmalade, milk, tea, and coffee. But the pleasures of the table depended greatly on the number of their prizes, for whenever possible they made every ship captured contribute heavily to their larder before sinking her. Of the tactics followed the observer writes:

> It appears that 55 per cent, or more than half, of the torpedoes fired miss their mark, and with this average they seem satisfied. Once they let go at a ship two torpedoes at 3000 yards' range, and both missed, the range being too long but they did not care to come any nearer, as they believed the ship to be well armed.
>
> They prefer to fire at 500 to 700 yards, which means that at this range the track or "wake" of a projectile would be discernible for, say, twenty-five to thirty seconds—not much time, indeed, for any ship to get out of the way. At 100 yards' range or less they do not care to fire unless compelled to, as the torpedo is nearly always discharged when the submarine is lying ahead of the object, *i. e.*, to hit the ship coming up to it; it follows that a gun forward is more useful than one aft, the gun aft being of real service when a submarine starts shelling, which she will do for choice from aft the ship rather than from forward of her, where she would be in danger of being run over and rammed.

CHAPTER 16

Submarine Warfare

At the moment of writing these words the outcome of the greatest war the world has ever known is believed by many to hang upon the success with which the Allies can meet and defeat the campaign of the German submarines. The German people believe this absolutely. The Allies and their sympathizers grudgingly admit that they are only too fearful that it may be true.

To such a marvellous degree of military efficiency has the ingenuity of man brought these boats which so recently as our Civil War were still in the vaguest experimental stage and scarcely possessed of any offensive power whatsoever!

Nevertheless these machines had reached a degree of development, and had demonstrated their dangerous character so early in the war that it was amazing that the British were so slow in comprehending the use that might be made of them in cutting off British commerce. It is true that the first submarine actions redounded in their results entirely to British credit. In September of 1914 a British submarine ran gallantly into Heligoland Bay and sank the German light cruiser *Hela* at her moorings. Shortly after the Germans sought retaliation by attacking a British squadron, but the effort miscarried. The British cruiser *Birmingham* caught a glimpse of her wake and with a well-aimed shot destroyed her periscope. The submarine dived, but shortly afterwards came up again making what was called a porpoise dive—that is to say, she came up just long enough for the officer in the conning tower to locate the enemy, then submerged again. Brief, however, as had been the appearance of the conning tower, the British put a shell into it and in a few minutes the submarine and most of her crew were at the bottom of the sea.

Soon after followed the attack upon and sinking of the three cruisers by the submarine under the command of Lieutenant Commander Otto von Weddigen, the narrative of which we have already told. But while after that attacks upon British armed ships were many, successes were few. There were no German ships at sea for the British to attack in turn, but some very gallant work was done by their submarines against Austrian and Turkish warships in the Mediterranean and the Dardanelles. All this time the Germans were preparing for that warfare upon the merchant shipping of all countries which at the end they came to believe would force the conclusion of the war. It seems curious that during this early period the Allies were able to devise no method of meeting this form of attack. When the United States entered the war more than three years later they looked to us for the instant invention of some effective anti-submarine weapon. If they were disappointed at our failure at once to produce one, they should have remembered at least that they too were baffled by the situation although it was presented to them long before it became part of our problems.

About no feature of the war have the belligerents thrown more of mystery than about the circumstances attending submarine attacks upon battleships and armed transports and the method employed of meeting them. Even when later in the war the Germans apparently driven to frenzy made special efforts to sink hospital and Red Cross ships the facts were concealed by the censors, and accounts of the efforts made to balk such inhuman and unchristian practices diligently suppressed. In the end it seemed that the British, who of course led all naval activities, had reached the conclusion that only by the maintenance of an enormous fleet of patrol boats could the submarines be kept in check. This method they have applied unremittingly. Alfred Noyes in a publication authorized by the British government has thus picturesquely told some of the incidents connected with this service:

> It is difficult to convey in words the wide sweep and subtle co-ordination of this ocean hunting; for the beginning of any tale may be known only to an admiral in a London office, the middle of it only to a commander at Kirkwall, and the end of it only to a trawler skipper off the coast of Ireland. But here and there it is possible to piece the fragments together into a complete adventure, as in the following record of a successful chase, where the glorious facts outrun all the imaginations of the wildest melodrama.

There were suspicious vessels at anchor, one moonless night, in a small bay near the Mumbles. They lay there like shadows, but before long they knew that the night was alive for a hundred miles with silent talk about them. At dawn His Majesty's trawlers *Golden Feather* and *Peggy Nutten* foamed up, but the shadows had disappeared.

The trawlers were ordered to search the coast thoroughly for any submarine stores that might have been left there. "Thoroughly" in this war means a great deal. It means that even the bottom of the sea must be searched. This was done by grapnels; but the bottom was rocky and seemed unfit for a base. Nothing was found but a battered old lobster pot, crammed with seaweed and little green crabs.

Probably these appearances were more than usually deceitful; for shortly afterward watchers on the coast reported a strange fishing boat, with patched brown sails, heading for the suspected bay. Before the patrols came up, however, she seemed to be alarmed. The brown sails were suddenly taken in; the disguised conning tower was revealed, and this innocent fishing boat, gracefully submerging, left only the smiling and spotless April seas to the bewildered eyes of the coast guard.

In the meantime signals were pulsing and flashing on land and sea, and the U-boat had hardly dipped when, over the smooth green swell, a great sea hawk came whirring up to join the hunt, a hawk with light yellow wings and a body of service grey—the latest type of seaplane. It was one of those oily seas in which a watcher from the air may follow a submarine for miles, as an olive green shadow under the lighter green. The U-boat doubled twice; but it was half an hour before her sunken shadow was lost to sight under choppy blue waters, and long before that time she was evidently at ease in her mind and pursuing a steady course. For the moment her trail was then lost, and the hawk, having reported her course, dropped out of the tale.

The next morning in the direction indicated by that report several patrol boats heard the sound of gunfire and overhauled a steamer which had been attacked by a submarine. They gave chase by "starring" to all the points of the compass, but could not locate the enemy. A little later, however, another trawler observed the wash of a submarine crossing her stern about two hundred yards away. The trawler star-boarded, got into the wake

A British Submarine

of the submarine and tried to ram her at full speed. She failed to do this, as the U-boat was at too great a depth. The enemy disappeared, and again the trawlers gathered and "starred."

In the meantime, certain nets had been shot, and, though the enclosed waters were very wide, it was quite certain that the submarine was contained within them. Some hours later another trawler heard firing and rushed toward the sound. About sunset she sighted a submarine which was just dipping. The trawler opened fire at once without result. The light was very bad and it was very difficult to trace the enemy, but the trawler continued the search, and about midnight she observed a small light close to the water. She steamed within a few yards of it and hailed, thinking it was a small boat. There was a considerable amount of wreckage about, which was afterward proved to be the remains of a patrol vessel sunk by the submarine. There was no reply to the hail, and the light instantly disappeared. For the third time the patrols gathered and "starred" from this new point.

And here the tale was taken up by a sailor who was in command of another trawler at the time. I give it, so far as possible, in his own words.

"About 4 o'clock in the morning I was called by Deckhand William Brown to come on deck and see if an object sighted was a submarine. I did so, and saw a submarine about a mile distant on the port bow. I gave the order, 'Hard a-starboard.' The ship was turned until the gun was able to bear on the submarine, and it was kept bearing. At the same time I ordered hands to station, and about ten minutes afterward I gave the order to fire. The submarine immediately altered her course from W. to N. N. W., and went away from us very fast. I burned lights to attract the attention of the drifters, and we followed at our utmost speed, making about eight knots and shipping light sprays. We fired another shot about two minutes later, but it was breaking dawn, and we were unable to see the fall of the shots. After the second shot the submarine submerged. I hoisted warning signals and about half an hour later I saw a large steamer turning round, distant between two and three miles on our starboard beam. I headed toward her, keeping the gun trained on her, as I expected, judging by her action, that she had smelt the submarine. When we were about a mile and a

SECTIONAL VIEW OF THE NAUTILUS

half from the steamer I saw the submarine half a mile astern of her. We opened fire again, and gave her four shots, with about two minutes between 'em. The submarine then dodged behind the off quarter of the steamer."

He paused to light his pipe, and added, quite gravely, "When she had disappeared behind the steamer I gave the order 'Cease fire,' to avoid hitting the larger vessel."

I made a mental note of his thoughtfulness; but, not for worlds would I have shown any doubt of his power to blast his way, if necessary, through all the wood and iron in the universe; and I was glad that the blue clouds of our smoke mingled for a moment between us.

"I saw two white boats off the port quarter," he continued. "But I paid no attention to them. I ordered the helm to be starboarded a bit more, and told the gunner to train his gun on the bow of the steamer; for I expected the submarine to show there next. A few minutes later she did so, and when she drew ahead I gave the order to fire. I should say we were about a mile and a quarter away. We gave him two more shots and they dropped very close, as the spray rose over his conning tower. He altered his course directly away from us, and we continued to fire. The third shot smothered his conning tower with spray. I did not see the fourth and fifth shots pitch. There was no splash visible, although it was then broad daylight; so I believe they must have hit him. A few moments after this the submarine disappeared.

"I turned, then, toward the two white boats and hailed them. The chief officer of the steamer was in charge of one. They were returning to their ship, and told me that we had hit the submarine. We escorted them through the nets and parted very good friends."

"But how did you get the scalp of this U-boat?" I asked.

"We signalled to the admiral, and sent the Daffy to investigate. She found the place, all right. It was a choppy sea, but there was one smooth patch in it, just where we told 'em the submarine had disappeared; a big patch of water like wavy satin, two or three hundred yards of it, coloured like the stripes on mackerel, all blue and green with oil. They took a specimen of the oil."

"Did it satisfy the Admiralty?"

"No. Nothing satisfies the Admiralty but certainties. They count the minimum losses of the enemy, and the maximum of their own. Very proper, too. Then you know where you are. But, mind you, I don't believe we finished him off that morning. Oil don't prove that. It only proves we hit him. I believe it was the *Maggie and Rose* that killed him, or the *Hawthorn*. No; it wasn't either. It was the *Loch Awe*."

"How was that?"

"Well, as Commander White was telling you, we'd shot out nets to the north and south of him. There were two or three hundred miles, perhaps, in which he might wriggle about; but he couldn't get out of the trap, even if he knew where to look for the danger. He tried to run for home, and that's what finished him. They'll tell you all about that on the *Loch Awe*."

So the next day I heard the end of the yarn from a sandy-haired skipper in a trawler whose old romantic name was dark with new significance. He was terribly logical. In his cabin—a comfortable room with a fine big stove—he had a picture of his wife and daughters, all very rigid and uncomfortable. He also had three books. They included neither Burns nor Scott. One was the *Bible*, thumbed by his grandfather and his father till the paper had worn yellow and thin at the sides. The second, I am sorry to say, was called *The Beautiful White Devil*. The third was an odd volume of Froude in the Everyman edition. It dealt with the Armada.

"I was towin' my nets wi' the rest o' my group," he said, "till about 3 o'clock i' the mornin' on yon occasion. It was fine weather wi' a kind o' haar. All at once, my ship gaed six points aff her coorse, frae S. E. to E. N. E., and I jaloused that the nets had been fouled by some muckle movin' body. I gave orders to pit the wheel hard a-port, but she wouldna answer. Suddenly the strain on the nets stoppit.

"I needna tell you what had happened. Of course, it was preceesely what the Admiralty had arranged tae happen when gentlemen in undersea boats try to cut their way through our nets. Mind ye, thae nets are verra expensive."

A different situation, however, has lately developed in the more unequal fight between submarines and merchant vessels. There the submarine unquestionably has gained and maintained supremacy. Two factors are primarily responsible for this: lack of speed and lack of armament on the part of the merchantman. Of course, recently the latter condition has been changed and apparently with good success. But even at best, an armed merchantman has a rather slim chance at escape. Neither space nor available equipment permits a general arming of merchantmen to a sufficient degree to make it possible for the latter to attack a submarine from any considerable distance. Then, too, what chance has a merchant vessel unprotected by patrol boats to escape the torpedo of a hidden submarine? How successfully this question will finally be solved, the future only will show. At present it bids fair to become one of the deciding factors in determining the final issue of this war.

The first authentically known case of an attack without warning by a German submarine against an allied merchantman was the torpedoing of the French steamship *Amiral Ganteaume* on October 26, 1914, in the English Channel. The steamer was sunk and thirty of its passengers and crew were lost. A number of other attacks followed during the remainder of 1914 and in January, 1915. Then came on February 3, 1915, the now famous pronouncement of the German Government declaring "all the waters around Great Britain and Ireland, including the whole of the English Channel, a war zone," and announcing that on and after Feb. 18th, Germany "will attempt to destroy every enemy ship found in that war zone, without its being always possible to avoid the danger that will thus threaten neutral persons and ships." Germany gave warning that "it cannot be responsible hereafter for the safety of crews, passengers, and cargoes of such ships," and it furthermore "calls the attention of neutrals to the fact that it would be well for their ships to avoid entering this zone, for, although the German naval forces are instructed to avoid all violence to neutral ships, in so far as these can be recognized, the order given by the British Government to hoist neutral flags and the contingencies of naval warfare might be the cause of these ships becoming the victims of an attack directed against the vessels of the enemy."

This was the beginning of the submarine controversy between Germany and the United States and resulted in a note from the United States Government in which it was stated that the latter viewed the possibilities created by the German note with such grave concern, that it feels it to be its privilege, and, indeed, its duty, in the circumstances to request the Imperial German Government to consider before action is taken the critical situation in respect of the relation between this country and Germany which might arise were the German naval forces, in carrying out the policy foreshadowed in the Admiralty's proclamation, to destroy any merchant vessel of the United States or cause the death of American citizens:—To declare and exercise a right to attack and destroy any vessel entering a prescribed area of the high seas without first certainly determining its belligerent nationality and the contraband character of its cargo would be an act so unprecedented in naval warfare that this Government is reluctant to believe that the Imperial Government of Germany in this case contemplates it as possible.

After stating that the destruction of American ships or American lives on the high seas would be difficult to reconcile with the friendly relations existing between the two Governments, the note adds that the United States "would be constrained to hold the Imperial Government of Germany to a strict accountability for such acts of their naval authorities, and to take any steps it might feel necessary to take to safeguard American lives and property and to secure to American citizens the full enjoyment of their acknowledged rights on the high seas."

It is not within the province of this book to go in detail into the diplomatic history of the submarine controversy between Germany and the United States. Suffice it to say, therefore, that from the very beginning the controversy held many possibilities of the disastrous ending which finally came to pass when diplomatic relations were broken off between the two countries on February 3, 1917, and a state of war was declared by President Wilson's proclamation of April 6, 1917.

The period between Germany's first War Zone Declaration and the President's proclamation—two months and three days more than two years—was crowded with incidents in which submarines and submarine warfare held the centre of the stage. It would be impossible within the compass of this story to give a complete survey of all the boats that were sunk and of all the lives that were lost. Nor would it be possible to recount all the deeds of heroism which this new warfare occasioned. Belligerents and neutrals alike were affected. American ships suffered, perhaps, to a lesser degree, than those of other neutrals, partly because of the determined stand taken by the United States Government. On May 1, 1915, the first American steamer, the *Gulflight*, was sunk. Six days later the world was shocked by the news that the *Lusitania*, one of the biggest British passenger liners, had been torpedoed without warning on May 7, 1915 and had been sunk with a loss of 1198 lives, of whom 124 were American citizens. Before this nation was goaded into war, more than 200 Americans were slain.

Notes were again exchanged between the two Governments. Though the German government at that time showed an inclination to

U. S. SUBMARINE *H-3* AGROUND ON CALIFORNIA COAST

abandon its position in the submarine controversy under certain conditions, sinkings of passenger and freight steamers without warning continued. All attempts on the part of the United States Government to come to an equitable understanding with Germany failed on account of the latter's refusal to give up submarine warfare, or at least those features of it which, though considered illegal and inhuman by the United States, seemed to be considered most essential by Germany.

Then came the German note of January 31, 1917, stating that "from February 1, 1917, sea traffic will be stopped with every available weapon and without further notice" in certain minutely described "prohibited zones around Great Britain, France, Italy, and in the Eastern Mediterranean."

The total tonnage sunk by German submarines from the beginning of the war up to February 1, 1917, has been given by British sources as over three million tons, while German authorities claimed four million. The result of the German edict for unrestricted submarine warfare has been rather appalling, even if it fell far short of German prophesies and hopes. During the first two weeks of February a total of ninety-seven ships with a tonnage of about 210,000 tons were sent to the bottom of the sea. Since then the German submarines have taken an even heavier toll. It has, however, become next to impossible, due to the restrictions of censorship, to compute any accurate figures for later totals, though it has become known from time to time that the Allied as well as the neutral losses have been very much higher during the five months of February to July, 1917 than during any other five months.

The figures of the losses of British merchantmen alone are shown by the following table:

Ships Week ending	1,600 Tons >	1,600 Tons <	Total
March 4	14	9	23
March 11	13	4	17
March 18	16	8	24
March 25	18	7	25
April 1	18	13	31
April 8	17	2	19
April 15	19	9	28
April 22	40	15	55
April 29	38	13	51
May 6	24	22	46
May 13	18	5	23

May	20	18	9	27
May	27	18	1	19
June	3	15	3	18
June	10	22	10	32
June	17	27	5	32
June	24	21	7	28
July	1	15	5	20
July	8	14	3	17
July	15	14	4	18
July	22	21	3	24
July	29	18	3	21
Aug.	5	21	2	23
Aug.	12	14	2	16
Aug.	19	15	3	18
Aug.	26	18	5	23
Sept.	2	20	3	23
Sept.	9	12	6	18
Sept.	16	8	20	28
Sept.	23	13	2	15
Sept.	30	11	2	13
Oct.	7	14	2	16
Oct.	14	12	6	18
Oct.	21	17	8	25
Oct.	28	14	4	18
Nov.	4	8	4	12
Nov.	11	1	5	6

The table with its week by week report of the British losses is of importance because at the time it was taken as a barometer indicative of German success or failure. The German admiralty at the moment of declaring the ruthless submarine war promised the people of Germany that they would sink a million tons a month and by so doing would force England to abject surrender in the face of starvation within three months. During that period the whole civilized world looked eagerly for the weekly statement of British losses. Only at one time was the German estimate of a million tons monthly obtained. Most of the time the execution done by the undersea boats amounted to less than half that figure. So far from England being beaten in three months, at the end of ten she was still unshattered, though sorely disturbed by the loss of so much shipping. Her new crops had come on and her statesmen declared that so far as the food supply was concerned they were safe for another year.

During this period of submarine activity the United States entered upon the war and its government immediately turned its attention to meeting the submarine menace. In the first four months literally nothing was accomplished toward this end. A few submarines were reported sunk by merchantmen, but in nearly every instance it was doubtful whether they were actually destroyed or merely submerged purposely in the face of a hostile fire. Americans were looked upon universally as a people of extraordinary inventive genius, and everywhere it was believed that by some sudden lucky thought an American would emerge from a laboratory equipped with a sovereign remedy for the submarine evil. Prominent inventors indeed declared their purpose of undertaking this search and went into retirement to study the problem. From that seclusion none had emerged with a solution at the end of ten months. When the submarine campaign was at its very height no one was able to suggest a better remedy for it than the building of cargo ships in such quantities that, sink as many as they might, the Germans would have to let enough slip through to sufficiently supply England with food and with the necessary munitions of war.

Many cruel sufferings befell seafaring people during the period of German ruthlessness on the high seas. An open boat, overcrowded with refugees, hastily provisioned as the ship to which it belonged was careening to its fate, and tossing on the open sea two or three hundred miles from shore in the icy nights of midwinter was no place of safety or of comfort. Yet the Germans so construed it, holding that when they gave passengers and crew of a ship time to take to the boats, they had fully complied with the international law providing that in the event of sinking a ship its people must first be given an opportunity to assure their safety.

There have been many harrowing stories of the experiences of survivors thus turned adrift. Under the auspices of the British government, Rudyard Kipling wrote a book detailing the agonies which the practice inflicted upon helpless human beings, including many women and children. Some of the survivors have told in graphic story the record of their actual experiences. Among these one of the most vivid is from the pen of a well-known American journalist, Floyd P. Gibbons, correspondent of the *Chicago Tribune*. He was saved from the British liner, *Laconia*, sunk by a German submarine, and thus tells the tale of his sufferings and final rescue:

> I have serious doubts whether this is a real story. I am not entirely certain that it is not all a dream and that in a few minutes I

will wake up back in stateroom B. 19 on the promenade deck of the Cunarder *Laconia* and hear my cockney steward informing me with an abundance of "and sirs" that it is a fine morning.

I am writing this within thirty minutes after stepping on the dock here in Queenstown from the British mine sweeper which picked up our open lifeboat after an eventful six hours of drifting, and darkness and baling and pulling on the oars and of straining aching eyes toward that empty, meaningless horizon in search of help. But, dream or fact, here it is:

The first-cabin passengers were gathered in the lounge Sunday evening, with the exception of the bridge fiends in the smoking-room. *Poor Butterfly* was dying wearily on the talking-machine and several couples were dancing.

About the tables in the smoke-room the conversation was limited to the announcement of bids and orders to the stewards. This group had about exhausted available discussion when the ship gave a sudden lurch sideways and forward. There was a muffled noise like the slamming of some large door at a good distance away. The slightness of the shock and the mildness of the report compared with my imagination was disappointing. Every man in the room was on his feet in an instant. I looked at my watch. It was 10.30.

Then came five blasts on the whistle. We rushed down the corridor leading from the smoking-room at the stern to the lounge, which was amidships. We were running, but there was no panic. The occupants of the lounge were just leaving by the forward doors as we entered.

It was dark when we reached the lower deck. I rushed into my stateroom, grabbed life preservers and overcoat and made my way to the upper deck on that same dark landing.

I saw the chief steward opening an electric switch box in the wall and turning on the switch. Instantly the boat decks were illuminated. That illumination saved lives.

The torpedo had hit us well astern on the starboard side and had missed the engines and the dynamos. I had not noticed the deck lights before. Throughout the voyage our decks had remained dark at night and all cabin portholes were clamped down and all windows covered with opaque paint.

The illumination of the upper deck, on which I stood, made the darkness of the water, sixty feet below, appear all the blacker when I peered over the edge at my station boat, No. 10.

Salvaging H-3, view I

Already the boat was loading up and men and boys were busy with the ropes. I started to help near a davit that seemed to be giving trouble, but was stoutly ordered to get out of the way and get into the boat. We were on the port side, practically opposite the engine well. Up and down the deck passengers and crew were donning lifebelts, throwing on overcoats, and taking positions in the boats. There were a number of women, but only one appeared hysterical....

The boat started downward with a jerk toward the seemingly hungry rising and falling swells. Then we stopped and remained suspended in mid-air while the men at the bow and the stern swore and tussled with the lowering ropes. The stern of the boat was down, the bow up, leaving us at an angle of about forty-five degrees. We clung to the seats to save ourselves from falling out.

"Who's got a knife? A knife! a knife!" bawled a sweating seaman in the bow.

"Great God! Give him a knife," bawled a half-dressed, gibbering negro stoker who wrung his hands in the stern.

A hatchet was thrust into my hand, and I forwarded it to the bow. There was a flash of sparks as it crashed down on the holding pulley. Many feet and hands pushed the boat from the side of the ship and we sagged down again, this time smacking squarely on the billowy top of a rising swell.

As we pulled away from the side of the ship its receding terrace of lights stretched upward. The ship was slowly turning over. We were opposite that part occupied by the engine rooms. There was a tangle of oars, spars and rigging on the seat and considerable confusion before four of the big sweeps could be manned on either side of the boat.

The gibbering bullet-headed negro was pulling directly behind me and I turned to quiet him as his frantic reaches with his oar were hitting me in the back.

"Get away from her, get away from her," he kept repeating. "When the water hits her hot boilers she'll blow up, and there's just tons and tons of shrapnel in the hold."

His excitement spread to other members of the crew in the boat. It was the give-way of nerve tension. It was bedlam and nightmare.

We rested on our oars, with all eyes on the still lighted *Laco-*

Salvaging H-3, view II

nia. The torpedo had struck at 10.30 P. M. It was thirty minutes afterward that another dull thud, which was accompanied by a noticeable drop in the hulk, told its story of the second torpedo that the submarine had despatched through the engine room and the boat's vitals from a distance of two hundred yards.

We watched silently during the next minute, as the tiers of lights dimmed slowly from white to yellow, then a red, and nothing was left but the murky mourning of the night, which hung over all like a pall.

A mean, cheese-coloured crescent of a moon revealed one horn above a ragged bundle of clouds low in the distance. A rim of blackness settled around our little world, relieved only by general leering stars in the zenith, and where the *Laconia*'s lights had shone there remained only the dim outlines of a blacker hulk standing out above the water like a jagged headland, silhouetted against the overcast sky.

The ship sank rapidly at the stern until at last its nose stood straight in the air. Then it slid silently down and out of sight like a piece of disappearing scenery in a panorama spectacle.

Boat No. 3 stood closest to the ship and rocked about in a perilous sea of clashing spars and wreckage. As our boat's crew steadied its head into the wind a black hulk, glistening wet and standing about eight feet above the surface of the water, approached slowly and came to a stop opposite the boat and not six feet from the side of it.

"What ship was dot?" The correct words in throaty English with a German accent came from the dark hulk, according to Chief Steward Ballyn's statement to me later.

"The *Laconia*," Ballyn answered.

"Vot?"

"The *Laconia*, Cunard Line," responded the steward.

"Vot did she weigh?" was the next question from the submarine.

"Eighteen thousand tons."

"Any passengers?"

"Seventy-three," replied Ballyn, "men, women, and children, some of them in this boat. She had over two hundred in the crew."

"Did she carry cargo?"

"Yes."

Salvaging H-3, view III

"Well, you'll be all right. The patrol will pick you up soon." And without further sound save for the almost silent fixing of the conning tower lid, the submarine moved off.

There was no assurance of an early pick-up, even tho the promise were from a German source, for the rest of the boats, whose occupants—if they felt and spoke like those in my boat—were more than mildly anxious about their plight and the prospects of rescue.

The fear of some of the boats crashing together produced a general inclination toward further separation on the part of all the little units of survivors, with the result that soon the small craft stretched out for several miles, all of them endeavouring to keep their heads in the wind.

And then we saw the first light—the first sign of help coming—the first searching glow of white brilliance, deep down on the sombre sides of the black pot of night that hung over us.

It was way over there—first a trembling quiver of silver against the blackness; then, drawing closer, it defined itself as a beckoning finger, altho still too far away yet to see our feeble efforts to attract it....

We pulled, pulled, lustily forgetting the strain and pain of innards torn and racked from pain, vomiting—oblivious of blistered hands and wet, half frozen feet.

Then a nodding of that finger of light—a happy, snapping, crap-shooting finger that seemed to say: "Come on, you men," like a dice-player wooing the bones—led us to believe that our lights had been seen. This was the fact, for immediately the coming vessel flashed on its green and red side-lights and we saw it was headed for our position.

"Come alongside port!" was megaphoned to us. And as fast as we could we swung under the stern, while a dozen flashlights blinked down to us and orders began to flow fast and thick.

A score of hands reached out, and we were suspended in the husky tattooed arms of those doughty British jack tars, looking up into the weather-beaten, youthful faces, mumbling thanks and thankfulness and reading in the gold lettering on their pancake hats the legend "H. M. S. *Laburnum*."

Of course, the submarine fleets of the various navies paid a heavy toll too. It has become, however, increasingly difficult to get any accurate figures of these losses. The British navy, it is known, has lost during 1914, 1915, and 1916 twelve boats, some of which foundered, were wrecked or mined while others simply never returned. The loss of eight German submarines has also been definitely established. Others, however, are known to have been lost, and their number has been greatly increased since the arming of merchantmen. In 1917 it was estimated that the Germans lost one U-boat a week and built three.

Just what sensations a man experiences in a submerged submarine that finds it impossible to rise again, is, of course, more or less of a mystery. For, though submarines, the entire crew of which perished, have been raised later, only one record has ever been known to have been made covering the period during which death by suffocation or drowning stared their occupants in the face. This heroic and pathetic record was written in form of a letter by the commander of a Japanese submarine, Lieutenant Takuma Faotomu, whose boat, with its entire crew, was lost on April 15, 1910, during manoeuvres in Hiroshima Bay. The letter reads in part as follows:

> Although there is, indeed, no excuse to make for the sinking of his Imperial Majesty's boat and for the doing away of subordinates through my heedlessness, all on the boat have discharged their duties well and in everything acted calmly until death. Although we are departing in pursuance of our duty to the State, the only regret we have is due to anxiety lest the men of the world may misunderstand the matter, and that thereby a blow may be given to the future development of submarines. While going through gasoline submarine exercise, we submerged too far, and when we attempted to shut the sluice-valve, the chain in the meantime gave way. Then we tried to close the sluice-valve, by hand, but it was too late, the rear part being full of water, and the boat sank at an angle of about twenty-five degrees.
>
> The switchboard being under water, the electric lights gave out. Offensive gas developed and respiration became difficult. The above has been written under the light of the conning-tower when it was 11.45 o'clock. We are now soaked by the water that has made its way in. Our clothes are very wet and we feel cold. I have always expected death whenever I left my home, and therefore my will is already in the drawer at Kara-

U. S. SUBMARINE *D 1* OFF WEEHAWKEN

saki. I beg, respectfully, to say to his Majesty that I respectfully request that none of the families left by my subordinates shall suffer. The only matter I am anxious about now is this. Atmospheric pressure is increasing, and I feel as if my tympanum were breaking. At 12.30 o'clock respiration is extraordinarily difficult. I am breathing gasoline. I am intoxicated with gasoline. It is 12.40 o'clock.

Could there be a more touching record of the way in which a brave man met death?

*****

More interest in submarine warfare than ever before was aroused in this country when the German war submarine *U-53* unexpectedly made its appearance in the harbour of Newport, R. I., during the afternoon of October 7, 1916. About three hours afterwards, without having taken on any supplies, and after explaining her presence by the desire of delivering a letter addressed to Count von Bernstorff, then German Ambassador at Washington, the *U-53* left as suddenly and mysteriously as she had appeared.

This was the first appearance of a foreign war submarine in an American port. It was claimed that the *U-53* had made the trip from Wilhelmshaven in seventeen days. She was 213 feet long, equipped with two guns, four torpedo tubes, and an exceptionally strong wireless outfit. Besides her commander, Captain Rose, she was manned by three officers and thirty-three men.

Early the next morning, October 8, it became evident what had brought the *U-53* to this side of the Atlantic. At the break of day, she made her re-appearance southeast of Nantucket. The American steamer *Kansan* of the American Hawaiian Company bound from New York by way of Boston to Genoa was stopped by her, but, after proving her nationality and neutral ownership was allowed to proceed. Five other steamships, three of them British, one Dutch, and one Norwegian were less fortunate. The British freighter *Strathend*, of 4321 tons was the first victim. Her crew were taken aboard the Nantucket shoals light-ship. Two other British freighters, *West Point* and *Stephano*, followed in short order to the bottom of the ocean. The crews of both were saved by United States torpedo boat destroyers who had come from Newport as soon as news of the *U-53*'s activities had been received there. This was also the case with the crews of the Dutch *Bloomersdijk* and the Norwegian tanker, *Christian Knudsen*.

Not often in recent years has there been put on American naval officers quite so disagreeable a restraint as duty enforced upon the commanders of the destroyers who watched the destruction of these friendly ships, almost within our own territorial waters, by an arrogant foreigner who gave himself no concern over the rescue of the crews of the sunken ships but seemed to think that the function of the American men of war. It was no secret at the time that sentiment in the Navy was strongly pro-Ally. Probably had it been wholly neutral the mind of any commander would have revolted at this spectacle of wanton destruction of property and callous indifference to human life. It is quite probable that had this event occurred before the invention of wireless telegraphy had robbed the navy commander at sea of all initiative, there might have happened off Nantucket something analogous to the famous action of Commodore Tatnall when with the cry, "Blood is thicker than water" he took a part of his crew to the aid of British vessels sorely pressed by the fire of certain Chinese forts on the Yellow River. As it was it is an open secret that one commander appealed by wireless to Washington for authority to intervene. He did not get it of course. No possible construction of international law could give us rights beyond the three-mile limit. He had at least however the satisfaction when the German commander asked him to move his ship to a point at which it would not interfere with the submarine's fire upon one of the doomed vessels, of telling him to move his own ship and accompanying the suggestion with certain phrases of elaboration thoroughly American.

The rapid development of submarine warfare naturally made it necessary to find ways and means to combat this new weapon of naval warfare. Much difficulty was experienced, especially in the beginning, because there were no precedents and because for a considerable period everything that was tried had necessarily to be of an experimental nature.

To protect harbours and bays was found comparatively easy. Nets were spread across their entrances. They were made of strong wire cables and to judge from the total absence of submarines within the harbours thus guarded they proved a successful deterrent. In most cases they were supported by extensive minefields. The danger of these to submarines, however, is rather a matter of doubt, for submarines can dive successfully under them and by careful navigating escape unharmed.

The general idea of fighting submarines with nets was also adopted for areas of open water which were suspected of being infested with submarines. Recently, serious doubts have been raised concerning the future usefulness of nets. Reports have been published that German submarines have been fitted up with a wire and cable cutting appliance which would make it possible for them to break through nets at will, supposing, of course, that they had been caught by the nets in such a way that no vital parts of the underwater craft had been seriously damaged. A sketch of this wire cutting device was made by the captain of a merchantman, who, while in a small boat after his ship had been torpedoed, had come close enough to the attacking submarine to make the necessary observations. The sketch showed an arrangement consisting of a number of strands of heavy steel hawsers which were stretched from bow to stern, passing through the conning tower and to which were attached a series of heavy circular knives a foot in diameter and placed about a yard apart. Even as early as January, 1915, Mr. Simon Lake, the famous American submarine engineer and inventor, published an article in the *Scientific American* in which he dwelt at length on means by which a submarine could escape mines and nets. One of the illustrations, accompanying this article, showed a device enabling submarines travelling on the bottom of the sea to lift a net with a pair of projecting arms and thus pass unharmed under it.

Many other devices to trap, sink or capture submarines have been invented. A large number of these, of course, have been found impracticable. Others, however, have been used with success. Few details of any of these have been allowed to become known.

The most dangerous power of submarines, is their ability to approach very closely to their object of attack without making their presence known to their prey. This naturally suggested that a way be found to detect the presence of submarines early enough to make it possible to stave off an attack or even to assume the offensive against the underwater boat. A recent invention, the perfection of which is due to the work of Mr. William Dubilier, an American electrical engineer, and of Professor Tissot, a member of the French Academy of Science, is the microphone. Few details are known about this instrument except that it records sound waves at as great a distance as fifty-five miles. This would permit in most cases the calling of patrol boats or the use of other defensive means before the submarine would be able to execute an attack.

At the present moment it would appear that the most dangerous enemy of the submarine yet discovered is the aeroplane or the dirigible. Some figures as to the mortality among submarines due

Submarine built for Spain in the Cape Cod canal

to the efforts of aircraft have been published in an earlier chapter. The chief value of aircraft in this work is due to the fact that objects under the water are readily discernible at a considerable depth when viewed from a point directly over them. An illustration familiar to every boy is to be found in the fact that he can see fish at the bottom of a clear stream from a bridge, while from the shore the refraction of the water is such that he can see nothing. From the air the aviator can readily see a submarine at a depth of fifty feet unless the water is unusually rough or turbid. The higher he rises the wider is his sphere of vision. With the lurking craft thus located the airman can either signal to watching destroyers or may bide his time and follow the submarine until it rises to the surface, when a well placed bomb will destroy it. Both of these methods have been adopted with success. For a time the submarines were immune from this form of attack because of the difficulty of finding a bomb which would not explode on striking the surface of the water, thus allowing its force to be dissipated before it reached the submarine, or else would not have its velocity so greatly checked by the water that on reaching the submarine the shock of its impact would not be great enough to explode it at all. Both of these difficulties have been overcome. The new high explosives have such power, taken in connection with the fact that water transmits the force of an explosion undiminished to a great distance, that many of them exploding at the surface will put out of action a submarine at a considerable depth. Furthermore bombs have been invented, which being fired, not merely dropped from an aeroplane, will go through the water with almost undiminished momentum and explode on striking the target, or after a period fixed by the assailant. Other bombs known as "depth bombs" are fitted with flanges that revolve as they sink, causing an explosion at any desired depth.

About the actual achievements of the aeroplane as a foe to submarines there hangs a haze of mystery. It has been the policy of the Allied governments to keep secret the record of submarines destroyed and particularly the methods of destruction. But we know that a few have met their fate from bolts dropped from the blue. In *The Outlook* Lawrence La Tourette Driggs, himself a flying man of no contemptible record, describes the method and result of such an attack. After recounting the steps by which a brother airman attained a position directly above a submerged submarine preparatory to dropping his bomb, he says:

Down shot his plummet of steel and neatly parted the waters ahead of the labouring submarine. But it did not explode. I could see a whirling metal propeller on the torpedo revolve as it sank. It must have missed the craft by twenty feet.

Suddenly a column of water higher than my position in the air stood straight up over the sea, then slipped noiselessly back. By all that is wonderful how did that happen?

As we covered the spot again and again in our circling machines, we were joined by two more pilots, and finally by a fast clipper steam yacht. The surface of the water was literally covered with oil, breaking up the ripple of the waves, and smoothing a huge area into gleaming bronze. Here and there floated a cork belt, odd bunches of cotton waste, a strip of carpet, and a wooden three-legged stool. These fragments alone remained to testify to the *corpus delicti*.

"Philip," I said half an hour later, as the hot coffee was thawing out our insides, "what kind of a civilized bomb do you call that?"

"That bears the simple little title of *trinitrotoluol*; call it T. N. T. for short," replied Sergeant Pieron.

"But what made it hang fire so long?" I demanded.

"It's made to work that way. When the bomb begins sinking the little propeller is turned as it is pulled down through the water. It continues turning until it screws to the end. There it touches the fuse-pin and that sets off the high explosive—at any depth you arrange it for."

I regarded him steadfastly. Then I remarked, "But it did not touch the submarine. I saw it miss."

"Yes, you can miss it fifty yards and still crush the submarine." He took up an empty egg shell. "The submarine is hollow like this. She is held rigidly on all her sides by the water. Water is non-compressible like steel. Now when the T. N. T. explodes, even some distance away, the violent expending concussion is communicated to this hollow shell just as though a battering ram struck it. The submarine can't give any because the surrounding water holds her in place. So she crumples up—like this."

Pieron opened his hand and the flakes of egg shell fluttered down until they struck the floor.

Gunfire undoubtedly is still the most reliable preventive against submarine attacks. Comparatively small calibred guns can cause serious

A CRITICAL MOMENT

damage to submarines even by one well directed shot. Submarines have been sunk both by warships and merchantmen in this way and many more have been forced to desist from attacks. Not every merchantman, of course, can be equipped with the necessary guns and gunners. Neither equipment nor men can be spared in sufficient quantities. But the efficiency of gun protection has been proved beyond all doubt by many authentic reports of successful encounters between armed merchantmen and submarines in which the latter were defeated.

Ramming, too, has been advocated and tried. It is, however, a procedure involving considerable danger to the attacking boat. For one thing all the submarine has to do is to dive quick and deep enough and it is out of harm's way. Then, too, the chances are that the submarine can launch a torpedo in time to reach the ramming vessel before the latter can do any damage.

There have been reports of submarine duels between Austrian and Italian submarines in the Adriatic in which it was claimed that in each at least one submarine was destroyed, and, at least, in one instance both the duellists were sunk. Generally speaking the fact has been established, however, that submarines cannot fight submarines with any degree of success, except in exceptional cases and under exceptional conditions.

Since the outbreak of the war between the United States and Germany the question of combating the submarine has become more acute than ever. The latest development has been along negative rather than affirmative lines. It has apparently been decided that none of the devices, known at present and capable of destroying submarines, is sufficient either alone or in combinations to defeat the submarines decisively. The best means of balancing as much as possible the losses which German submarines are inflicting on the shipping facilities of the Allies at the present seems to be the unlimited and prompt building of large fleets of comparatively small ships. If this can be accomplished in time, the German submarines undoubtedly will find it impossible to destroy a tonnage sufficient to exert any great influence on the final outcome of the war.

CHAPTER 17

The Future of the Submarine

The world will not always be at war. Interminable as the conflict by which it is now racked seems, and endless as appear the resources of the nations participating in it, the time must come when victory or sheer exhaustion shall compel peace. People talk of that peace being permanent. That is perhaps too sanguine a dream while human nature remains what it is, and nations can still be as covetous, ambitious, and heedless of others' rights as are individuals. But beyond doubt a prolonged period of peace awaits the world. What then is to be the future of the aircraft and the submarine which had to wait for war to secure any recognition from mankind of their prodigious possibilities?

Of the future of the aircraft there can be no doubt. Its uses in peace will be innumerable. Poor old Count Zeppelin, who thought of his invention only as a weapon of war, nevertheless showed how it might be successfully adapted to the needs of peace merely as a by-product. As for the aeroplane both for sport and business its opportunities are endless. Easy and inexpensive to build, simple to operate with but little training on the part of the aviator, it will be made the common carrier of all nations. Already the United States is maintaining an aerial mail service in Alaska. Already too, bi- and triplanes are built capable of carrying twenty-five to thirty men besides guns and ammunition. It is easy to foresee the use that can be made of machines of this character in times of peace. Needing no tracks or right of way, requiring no expensive signalling or operative system, asking only that at each end of the route there shall be a huge level field for rising and for landing, these machines will in time take to themselves the passenger business of the world.

But the future of the submarine is more dubious. Always it will be a potent weapon of war. It may indeed force the relegation of dreadnoughts to the scrap heap. But of its peaceful services there is more doubt. That it can be made a cargo carrier is unquestionably true. But to what good? There is no intelligent reason for carrying cargoes slowly under water which might just as well be carried swiftly on the surface unless war compels concealment. Underwater navigation must always be slower and more expensive than surface navigation, nor does it seem probable that the underwater boats can ever equal in size ordinary ships, though undoubtedly their present proportions are going to be greatly increased.

As a result of the German submarine campaign it is possible that the United States may develop a fleet of underwater merchantmen to circumvent the enemy while this war continues, though there has been but little discussion of it. But even so, commonsense would indicate that such a fleet would be abandoned on the restoration of peace. If anything is to be done toward making the submarine a vessel of ordinary everyday use the present double system of motors—the Diesels for surface navigation and the electric for submerged service—will have to be abandoned. Inventors however are diligently working on this problem to-day. Indeed so well known and successful a builder of submarines as Mr. Simon Lake seemed to have faith in their possibilities as merchant craft. As early as February, 1916, he announced that he had taken out a patent on a new form of cargo-carrying submarine which he described as made up of "nests of light-weight circular tanks of comparatively small diameter surrounded by a ship-shape form of hull." What advantage was to accrue from this type of vessel Mr. Lake has not explained. However the Germans who seemed to originate everything successfully demonstrated that the merchant submarine was a practicable and useful craft with which to beat the blockade.

This was proved by the two successful trips made by the unarmed German merchant submarine *Deutschland* between Germany and the United States in 1916. Loaded with a cargo of dyestuffs and chemicals she left Bremen on June 14, 1916, and arrived in Baltimore early in July. After a short stay, during which she took on a full return cargo, consisting chiefly of rubber and metal, she started on August 1, 1916, for her return trip to Bremen where she arrived safely soon after August 15, 1916. Once more, in October of the same year she made a successful round trip, docking this time in New London. There was considerable talk about additional trips by other German merchant submarines, but none of them were ever carried out. It has never

A SUBMARINE BUILT FOR CHILE, PASSING THROUGH CAPE COD CANAL

become known whether this was due to the loss of these merchant submarines or to political relations between Germany and the United States which were then gradually assuming a less friendly form.

Of course, it is true that such boats are blockade runners and in a way, therefore, part and parcel of warfare. But they are unarmed merchantmen just the same and their exclusively mercantile character has been officially acknowledged by the United States Government. Under conditions of peace, however, it is very doubtful whether submarine merchantmen would pay, nor does it seem as if they possessed any advantages at all over surface merchant vessels. Nevertheless they represent an entirely new development of submarine navigation and, therefore, deserve attention.

During her stay in the United States, very few people were permitted to get more than a glance of the *Deutschland*. As a result, comparatively little became known regarding her mechanical details. The *Scientific American*, however, in its issue of July 22, 1916, gives a fairly detailed description of this first merchant submarine.

From this account we learn that the *Deutschland* conforms rather closely to the typical German naval U-boat. The hull proper consists of an internal cigar-shaped, cylindrical structure, which extends from stem to stern, and in its largest diameter measures about twenty feet. Enclosing this hull is a lighter false hull, which is perforated, to permit the entrance and exit of the sea-water, and is so shaped as to give the submarine a fairly good ship model for driving at high speed on the surface and at a much lesser speed submerged. The upper portion of the false hull does not present such a flat deck-like appearance as is noticeable in the naval U-boats. In fact, the whole modelling of the *Deutschland*, as compared with the naval boats, suggests that she has been filled out somewhat, with a view to obtaining the necessary displacement for cargo carrying.

The interior cylindrical hull is divided by four transverse bulkheads into five separate water-tight compartments. Compartment No. 1, at the bow, contains the anchor cables and electric winches for handling the anchor; also general ship stores, and a certain amount of cargo. Compartment No. 2 is given up entirely to cargo. Compartment No. 3, which is considerably larger than any of the others, contains the living quarters of the officers and crew. At the after end of this compartment, and communicating with it, is the conning tower. Compartment No. 4 is given up entirely to cargo. Compartment No. 5 contains the propelling machinery, consisting of two heavy oil engines and two electric motors. The storage batteries are carried in the bottom of the boat, below the living compartment. For purposes of communication, a gangway, 2 feet 6 inches wide by 6 feet high, is built through each cargo compartment, thus rendering it possible for the crew to pass entirely from one end of the boat to the other.

The length of the *Deutschland* is about 315 feet; beam 30 feet, and draught 17 feet. For surface propulsion and for charging the batteries, the boat carries two 4-cylinder, Diesel, heavy-oil motors of about 600 H. P. each. The speed at the surface is from 12 to 13 knots; and submerged it is 7 knots. At the surface the displacement of the boat is about 2000 tons, and she has a cargo capacity of about 700 tons.

The freeboard to the main deck, which runs the full length of the boat, but is only about 5½ feet wide, is about 6 feet, and the cockpit at the top of the conning tower is about 15 feet above the water. This cockpit, by the way, is suggestive of the protection afforded a chauffeur in an automobile, there being a shield in front of the quartermaster, so shaped as to throw the wind and spray upwards and clear of his face.

Two periscopes are provided; one at the forward end of the conning tower, and the other, of larger diameter, being forward and on the starboard of the conning tower. An interesting feature is the two folding, steel, wireless masts, about 50 feet in height, both of which fold aft into pockets built in the deck of the ship. The forward one of these masts carries a crow's nest for the lookout.

The commander of the *Deutschland*, Captain Paul König, was before the war a popular captain of North German Lloyd liners. He has published a very vivid and interesting account of the *Deutschland*'s trip, the *Voyage of the Deutschland*. In this book, he tells us how he was offered this novel command while the plans were still being drawn and that he immediately accepted, making, however, the proviso "if the thing really comes off."

The men, backing the venture, lost no time and, so Captain König tells us—

> .... in less than two months a telegram called me to Berlin to an important conference. Here I looked at sketches, plans, and working drawings until my eyes swam. Four more months passed which I utilized to the full. I then went to Kiel and saw a remarkable framework of steel slowly take shape upon the stocks across the way at Gaarden. Rotund, snug, and harmless the thing lay there. Inside it were hidden all the countless, complicated,

and powerful features of those sketches and working drawings. I cannot boast that the reality as executed in steel and brass was any easier to grasp than the endless network of lines and circles which had bewildered me when inspecting the blueprints.

Those of you who have seen illustrations and photographs of the interior of the "central station" or the "turret" of a submarine, will understand what I mean. And should you have entered a submarine itself and felt yourself hopelessly confused by the bewildering chaos of wheels, vents, screws, cocks, pipes, conduits—above, below, and all about—not to speak of the mysterious levers and weird mechanisms, each of which has some important function to fulfil, you may find some consolation in the thought that my own brains performed a devils' dance at the sight.

But after this monster, with its tangle of tubes and pipes, had been duly christened, and its huge grey-green body had slid majestically into the water, it suddenly became a ship. It swam in its element as though born to it—as though it had never known another.

For the first time I trod the tiny deck and mounted the turret to the navigation platform. From here I glanced down and was surprised to see beneath me a long, slender craft—with gracious lines and dainty contours. Only the sides, where the green body vaulted massively above the water, gave an indication of the huge size of the hull. I felt pride and rapture as my eye took in this picture. The fabric swayed slightly beneath my feet—an impressive combination of power and delicacy.

And now I know that what had at first seemed to me nothing more than the product of some mad fantasy on the part of the technicians was in reality a ship. It was a ship in which oceans might be crossed, a real ship, to which the heart of an old sailor like myself might safely attach itself.

Then came a short period of trial trips and diving tests, all of which were carried off successfully, and at last the day of departure arrived. As soon as the last escort had turned around a final diving test was ordered.

Instantly the response came back from the turret and the central station, and the men hurried to their posts. The oil engines were still hammering away at a mad rate. I left the manhole of the turret. The cover was battened down, the engines stopped at the same moment.

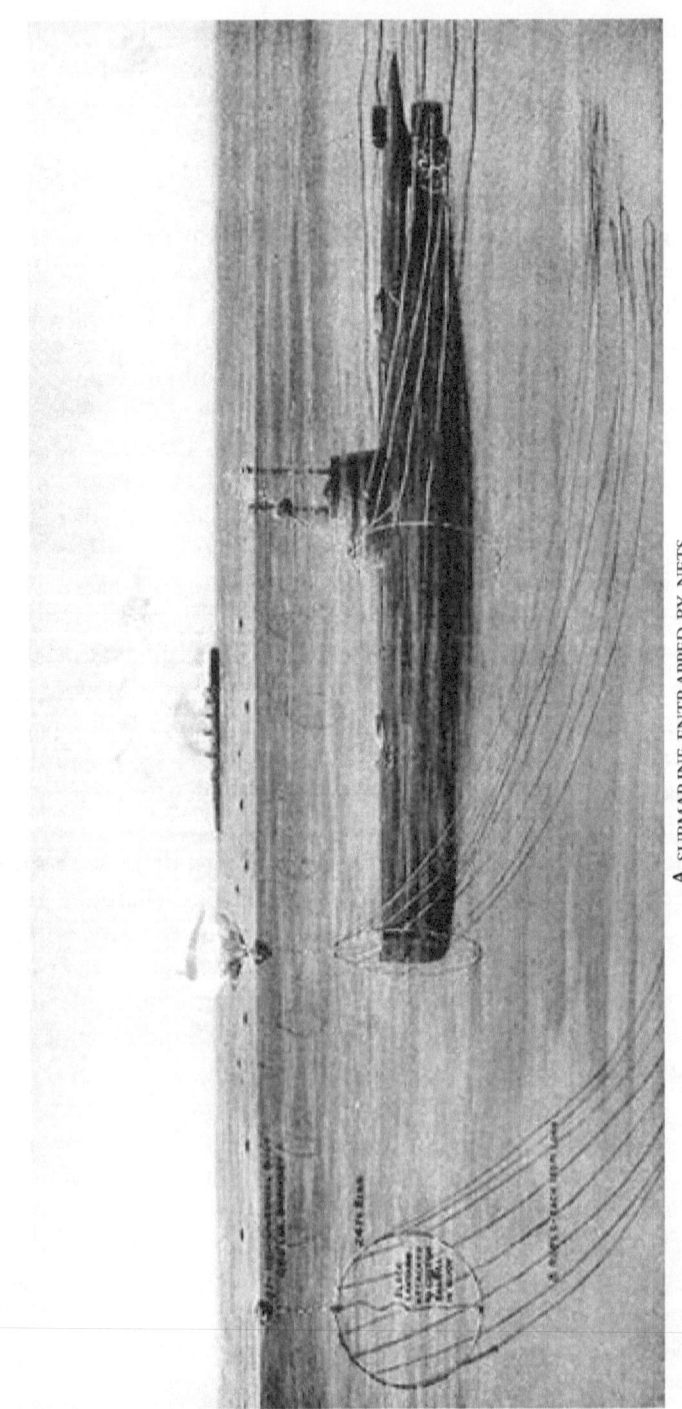

A SUBMARINE ENTRAPPED BY NETS

We felt a slight pressure in our ears for a moment. We were cut off from outside and silence reigned. But this silence was merely an illusion—and was due to the change.

"Open the diving-valves! Submerge!"

The valves were flung open and the compressed air escaped hissing from the tanks. At the same time a gigantic, intermittent snorting ensued, like the blowing and belching of some prehistoric monster. There was an uncomfortable pressure in our ears, then the noise became more regular, followed by a buzzing and a shrill hum. All the high notes of the engines in the central station intermingled and made a bewildering noise. It was like a mad diabolical singsong. And yet it was almost like silence after the dull, heavy pounding of the oil-motors—only more insistent and irritating. The penetrating hum in the various vents announced the fact that the diving mechanism was in operation. It moaned and sang lower and lower in the scale of tones. These slowly diminishing and steadily deepening tones give one the physical feeling of mighty volumes of water pouring in and flooding full.

You have the sensation of growing heavier and sinking as the boat grows heavier and sinks, even though you may not be able to see through the turret window, or the periscope, how the bows are gradually submerged and the water climbs higher and higher up the turret until all things without are wrapped in the eerie twilight of the depths.

The faithful lamps burned, however, and then a real silence suddenly ensued. There was no sound but the gentle trembling rhythm of the electric engines.

I then gave the order:

"Submerge to twenty meters!"

"Both engines half steam ahead!"

I was able to follow our submersion by means of the manometer. Through flooding the tanks, the boat is given several tons over-weight and the enclosed ship's space is made heavier than the displaced quantity of water. The titanic fish, therefore, began to sink downward in its element, that is to say, it began, in a certain sense, to fall. At the same time the electric engines are put into motion and the propulsive force of the propellers acts upon the diving rudders and causes the sinking to become a gliding. After the required depth has been reached—something which may easily be read from the manometer that records the depth—all further sinking may be stopped by simply lightening the hull, which is done by forcing out some of the water in the submarine's tanks. The furious growling of the pump is always a sure sign that the required depth is being approached. The noise ceased, only the electric motors continued to purr, and the word came from the central station:

"Twenty meters—even keel!"

"Rudder set!"

So we forged ahead at a depth of twenty meters. Of course we are "blind" under such conditions and can regulate our movements only by means of the depth recorder and that precious little jewel of the boat, our compass. No ray of light reached us any longer from without, the periscope was submerged long ago and the steel safety covers over the windows were closed. We had been metamorphosed completely into a fish.

Orders were then given to rise again. The *Deutschland* carried out this manoeuvre with the same facility with which she had taken the initial dive of her long voyage. In record time the ballast tanks were emptied and the change from electric motors to oil engines was completed without further loss of time. The boat was started at top surface speed towards her ultimate goal, the United States.

On the following day the *Deutschland* barely escaped running foul of a British submarine chaser, disguised as a neutral merchantman. A quick dive alone saved her. When she came up again a wild storm and a heavy sea were raging. Even before the change from the electric motors to the oil engines had been completed, another dangerous looking vessel appeared and before long was recognized as a hostile destroyer by Captain König. He tells us that he "Made one jump into the turret and slammed the cover fast."

"Alarm! Dive quickly! Flood!"

"Set diving rudder!"

"Twenty meters' depth!"

The commands were uttered in almost one breath. But the execution of them!

To attempt to dive with such a sea running was sheer madness, as experience has taught us. What was I to do? The destroyer might have seen us already!

Well, we knew we must get under—and as quickly as possible.

The men in the central below me were working away in silent haste. All the exhausts were opened wide, the compressed air hissed from the tanks—the diving vents were chanting in all possible keys.

I stood with my lips pressed together and stared out of the turret window upon the tossing sea, and watched for the first sign of our going down. But our deck remained still visible and we were continually lifted into the air by some wave. There was not a moment to be lost.

I ordered the diving rudder to be set still more sharply and both engines to drive ahead with full power.

The whole vessel quivered and thrilled under the increased pressure of the engines and made several leaps. She staggered about in the furious seas—but still seemed loath to leave the surface. Then she gave a jerk and her bows suddenly dipped and cut into the flood. She began to sink into the depths at an ever-increasing angle. The coming daylight vanished from the windows of the turret, the manometer in rapid succession showed 2—3—6—10 meters' depth. But the angle of the boat also began to increase.

We staggered about, leaned back, slipped off our feet. We then lost our footing entirely—for the floor of the *Deutschland* slanted sharply toward the front. I was just able to catch hold of the ocular or eye-piece of the periscope. Down in the central the men were hanging on to the hand-wheels of the diving rudder. A few terrible seconds passed thus.

We had not yet seized the full significance of this new situation when there came a severe shock. We were hurled to the floor and everything that was not fastened down went flying in all directions.

We found ourselves in the queerest attitudes—and stared into one another's faces. There was a grim silence for a moment, then First Officer Krapohl remarked dryly:

"Well, we seem to have arrived!"

This broke the ghastly tension.

We were all rather pale around the gills, but at once tried to get our bearings.

What had happened?

What had caused this unnatural inclination of the boat? And

Diagram of a German submarine mine-layer captured by British

why were the engines above us raving at intervals in a way that made the whole boat roar from stem to stern?

Before any of us had arrived at any solution of the mystery, our Chief Engineer, little Klees, had jumped up from his crouching position, and, swift as lightning, had swept the engine-signal dial around to "Stop!"

And suddenly there was a deep silence.

We slowly assembled our proper legs and arms and thought hard over what had happened.

The vessel had slanted down toward the bows at an angle of about 36 degrees. She was standing, so to speak, on her head. Our bow was fast upon the bottom of the sea—our stern was still oscillating up and down like a mighty pendulum. The manometer showed a depth of about 15 meters.

However, the *Deutschland* finally worked herself free and soon was again on the surface. Luck must have been with her, for she had suffered no damage and, in spite of the mountains of water which she must have thrown up, the hostile destroyer had not discovered her. Once more she was off on her way.

So the days went by and before long the merchant submarine had passed, without having been detected, beyond the territory in which British patrol boats were operating. Then came a succession of uneventful days and fine weather. Practically every day diving tests were made. One of these the captain describes as follows:

During these experimental diving tests we were treated to a spectacle of fairy-like loveliness.

I had set the rudder in such a way that the turret was travelling about three yards under water. Overhead the sun shone brilliantly and filled the deeps with a clear radiance. The pure water was luminous with colour—close at hand it was of a light azure blue, of fabulous clearness and transparent as glass. I could see the entire boat from the turret windows. The shimmering pearls of the air-bubbles which rise constantly from the body of the craft played about the entire length of the vessel from deck to bows, and every detail stood out in miraculous sharpness. Farther ahead there was a multi-coloured twilight. It seemed as if the prow kept pushing itself noiselessly into a wall of opalescent green which parted, glistening, and grew to an ethereal, rainbow-like translucency close at hand.

We were spell-bound by this vision of beauty. The fairy-like effect was increased by medusa which, poised in the transparent blue, frequently became entangled in the wires of the mine-guards or the railings and glowed like trembling fires of rose, pale gold, and purple.

But less pleasant things were in store for the *Deutschland*'s crew. The nearer the boat came to the region of the Gulf Stream, the more violent the weather became. Though she still ran most of the time on the surface, it became necessary to keep all openings battened down. Even the manhole, leading to the turret, could be kept open only for short periods. Naturally the temperature was rising all the time. It was midsummer and the Gulf Stream contributed its share of warmth. No wonder, therefore, that Captain König compares conditions below decks to a "veritable hell," and then continues:

While in the Gulf Stream we had an outer temperature of 28° Celsius. This was about the warmth of the surrounding water. Fresh air no longer entered. In the engine-room two 6-cylinder combustion motors kept hammering away in a maddening two-four time. They hurled the power of their explosions into the whirling crankshafts. The red-hot breath of the consumed gases went crashing out through the exhausts, but the glow of these incessant firings remained in the cylinders and communicated itself to the entire oil-dripping environment of steel. A choking cloud of heat and oily vapour streamed from the engines and spread itself like a leaden pressure through the entire ship.

During these days the temperature mounted to 53° Celsius.

And yet men lived and worked in a hell such as this! The watch off duty, naked to the skin, groaned and writhed in their bunks. It was no longer possible to think of sleep. And when one of the men fell into a dull stupor, then he would be aroused by the sweat which ran incessantly over his forehead and into his eyes, and would awake to new torment.

It was almost like a blessed deliverance when the eight hours of rest were over, and a new watch was called to the central or the engine-room.

But there the real martyrdom began. Clad only in an under-shirt and drawers, the men stood at their posts, a cloth wound about their foreheads to keep the running sweat from streaming into their eyes. Their blood hammered and raced in their tem-

A SUBMARINE DISCHARGING A TORPEDO

ples. Every vein boiled as with fever. It was only by the exertion of the most tremendous willpower that it was possible to force the dripping human body to perform its mechanical duty and to remain upright during the four hours of the watch....

But how long would we be able to endure this?

I no longer kept a log during these days and I find merely this one note: "Temperature must not rise any higher if the men are to remain any longer in the engine-room."

But they did endure it. They remained erect like so many heroes, they did their duty, exhausted, glowing hot, and bathed in sweat, until the storm centre lay behind us, until the weather cleared, until the sun broke through the clouds, and the diminishing seas permitted us once more to open the hatches.

The *Deutschland* was now near her goal. Without any trouble she entered Hampton Roads and was docked at Baltimore. There her cargo was discharged and her return cargo loaded. This latter operation involved many difficulties. During her stay a United States Government Commission made a detailed inspection of the *Deutschland* to determine beyond all question her mercantile character. But at last the day of departure, August 1st, had arrived. Properly escorted she made the trip down the Patapsco River and Chesapeake Bay. On her way down she made again diving trials which Captain König describes as follows:

> In order to see that everything else was tight and in good order, I gave the command to set the boat upon the sea bottom at a spot which, according to the reading upon the chart, had a depth of some 30 meters.
>
> Once again everything grew silent. The daylight vanished the well-known singing and boiling noise of the submerging vents vibrated about us. In my turret I fixed my eyes upon the manometer. Twenty meters were recorded, then twenty-five. The water ballast was diminished—thirty meters appeared and I waited the slight bump which was to announce the arrival of the boat at the bottom.
>
> Nothing of the sort happened.
>
> Instead of this the indicator upon the dial pointed to 32—to 33—to 35 meters....
>
> I knocked against the glass with my finger—correct—the arrow was just pointing toward thirty-six.

"Great thunder! what's up?" I cried, and reached for the chart. Everything tallied. Thirty meters were indicated at this spot and our reckoning had been most exact.

And we continued to sink deeper and deeper.

The dial was now announcing 40 meters.

This was a bit too much for me. I called down to the central and got back the comforting answer that the large manometer was also indicating a depth of over forty meters!

The two manometers agreed.

This, however, did not prevent the boat from continuing to sink.

The men in the central began to look at one another....

Ugh! it gives one a creepy feeling to go slipping away into the unknown amidst this infernal singing silence and to see nothing but the climbing down of the confounded indicator upon the white-faced dial....

There was nothing else to be seen in my turret. I glanced at the chart and then at the manometer in a pretty helpless fashion.

In the meantime the boat sank deeper; forty-five meters were passed—the pointer indicated forty-eight meters. I began to think the depth of the Chesapeake Bay must have some limit; we surely could not be heading for the bottomless pit? Then—the boat halted at a depth of fifty meters without the slightest shock.

I climbed down into the central and took counsel with Klees and the two officers of the watch.

There could be only one explanation; we must have sunk into a hole which had not been marked upon the chart.

When orders were now given to rise, it was found that the exhaust pumps refused to work. After a while, however, the chief engineer succeeded in getting them started. They reached the surface after about two hours of submergence.

It was dark by the time the merchant submarine was approaching the three-mile limit. Outside of it hostile warships were lying in wait. That the *Deutschland* escaped them well illustrates the fact that submarines may be kept by various means from entering a bay or a harbour, but that to blockade their exit is practically impossible. This is how Captain König speaks of his escape.

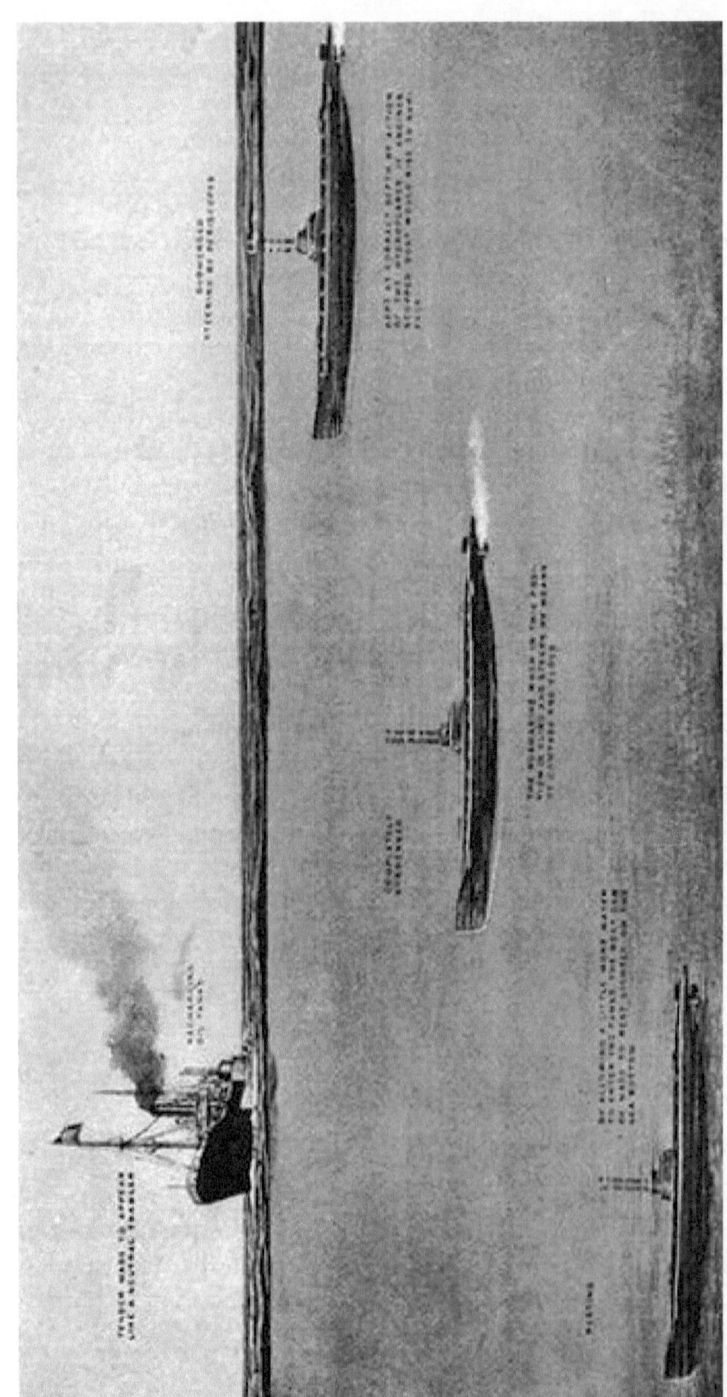

A German Submarine in Three Positions

We knew that the most dangerous moment of our entire voyage was now approaching. We once more marked our exact position, and then proceeded to make all the preparations necessary for our breaking through.

Then we dived and drove forward. All our senses were keyed to the utmost, our nerves taut to the breaking-point with that cold excitement which sends quivers through one's soul, the while outwardly one remains quite serene, governed by that clear and icy deliberation which is apt to possess a man who is fully conscious of the unknown perils toward which he goes....

We knew our path. We had already been informed that fishermen had been hired to spread their nets along certain stretches of the three-mile limit; nets in which we were supposed to entangle ourselves; nets into which devilish mines had very likely been woven....

Possibly these nets were merely attached to buoys which we were then supposed to drag along after us, thus betraying our position....

We were prepared for all emergencies, so that in case of extreme necessity we should be able to free ourselves of the nets. But all went well.

It was a dark night. Quietly and peacefully the lighthouses upon the two capes sent forth their light, the while a few miles further out death lay lowering for us in every imaginable form.

But while the English ships were racing up and down, jerking their searchlights across the waters and searching again and again in every imaginable spot, they little surmised that, at times within the radius of their own shadows, a periscope pursued its silent way, and under this periscope the U-*Deutschland*.

That night at twelve o'clock, after hours of indescribable tension, I gave the command to rise.

We Had Broken Through!

Slowly the *Deutschland* rose to the surface, the tanks were blown out and the Diesel engines flung into the gearing. At our highest speed we now went rushing toward the free Atlantic.

The homeward voyage was completed without untoward incident and long before the month had ended, the first—and probably last—merchant submarine was again safe and snug in her home port.

The cargo-carrying submarine, however, is by no means the only type of underwater vessel engaged in peaceful pursuits which has been suggested so far. Mr. Simon Lake, the American submarine engineer and inventor, has frequently pointed out the commercial possibilities of the submarine.

In the early part of 1916 a series of articles from his pen appeared in *International Marine Engineering*. They contained a number of apparently feasible suggestions looking towards the commercial development of the submarine.

First of all he tells of experiments made with submarines for navigation under ice. The proper development of this idea, of course, would be of immense commercial value. Many harbours in various parts of the world are inaccessible during the winter months for vessels navigating on the surface. Navigation on many important inland lakes likewise has to be stopped during that period. Submarines, built so that they can safely travel under the ice, would overcome these conditions and would make it possible to use most ice-bound ports throughout the entire year at least in Mr. Lake's view.

Ever since Mr. Lake began inventing and building submarines he has been interested in the possibilities which submarines offer for the exploration of the sea-bottom and for the discovery of wrecks and recovery of their valuable cargoes. His first boat, the *Argonaut*, as we have heard, possessed a diving chamber for just such purposes. He has continued his investigations and experiments along this line, and in these articles he shows illustrations of submarine boats and devices adapted for such work. Properly financed and directed, the recovery of cargoes from wrecks undoubtedly would not only bring large financial returns to the backers of such a venture, but also do away with the immense waste which the total loss of sunken vessels and cargoes inflicts now on the world. Submarines in peace may yet recover for the use of man much of the wealth which submarines in war have sent to the bottom of the sea. Marine insurance, too, would be favourably affected by such an undertaking.

Still one other commercial submarine boat is advocated by Mr. Lake. This is to be used for the location and collection of shellfish on a large scale. Of this vessel its inventor says:

The design of this submarine oyster-dredging vessel is such that the vessel goes down to the bottom direct, and the water is forced out of the centre raking compartment so that the oysters may be seen by the operator in the control compartment. With only a few inches of water over them, headway is then given to the submarine

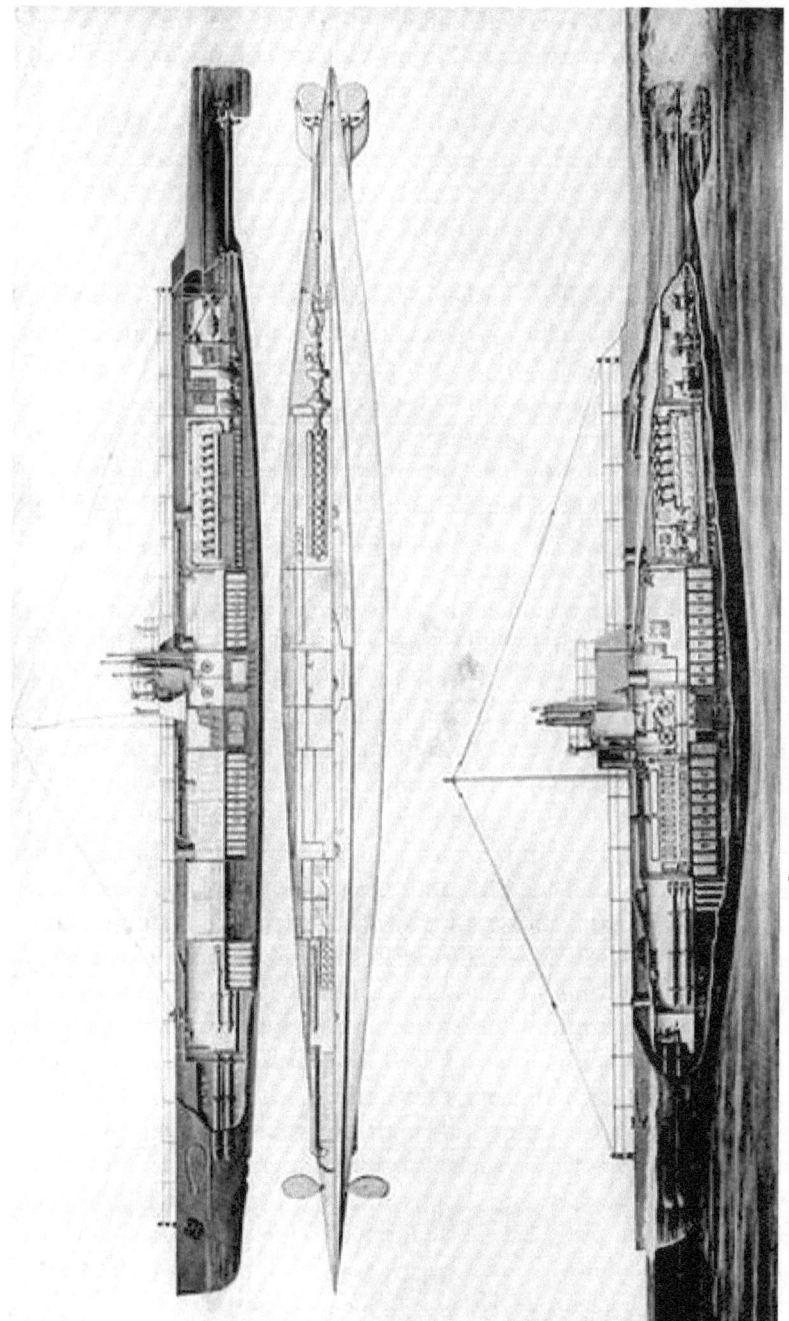

SECTIONAL VIEW OF A BRITISH SUBMARINE

and the oysters are automatically raked up, washed, and delivered through pipes into the cargo-carrying chambers. Centrifugal pumps are constantly delivering water from the cargo compartments, which induces a flow of water through the pipes leading from the "rake pans" with sufficient velocity to carry up the oysters and deposit them into the cargo holds. In this manner the bottom may be seen, and by "tracking" back and forth over the bottom the ground may be "cleaned up" at one operation.

This boat has a capacity of gathering oysters from good ground at the rate of five thousand bushels per hour. The use of the submarine will make the collection of oysters more nearly like the method of reaping a field of grain, where one "swathe" systematically joins on to another, and the whole field is "cleaned up" at one operation.

Man's greediness for profit has already driven the salmon from the rivers of New England where once they swarmed. Mechanical devices for taking them by the hundreds of thousands threaten a like result in the now teeming rivers of Washington and British Columbia. Mr. Lake's invention has the demerit of giving conscienceless profiteers the opportunity to obliterate the oyster from our national waters.

It does not appear, however, that, except as an engine of war the submarine offers much prospect of future development or future usefulness. And as we of the United States entered this war, which now engages our energies and our thoughts, for the purpose of making it the last war the world shall ever know, speculation on the future of the submarine seems rather barren. That does not mean however that there will be a complete stoppage of submarine construction or submarine development. War is not going to be ended by complete international disarmament, any more than complete unpreparedness kept the United States out of the struggle. A reasonable armament for every nation, and the union of all nations against any one or two that threaten wantonly to break the peace is the most promising plan intelligent pacifism has yet suggested. In such an international system there will be room and plenty for submarines.

Indeed it is into just such a plan that they intelligently fit. Though not wholly successful in their operations against capital ships, they have demonstrated enough power to make nations hesitate henceforth before putting a score of millions into ponderous dreadnoughts which have to retire from submarine-infested waters as the British did in their very hour of triumph at Jutland. They have not nullified, but greatly reduced the value of overwhelming sea power such as the

British have possessed. A navy greater than those of any two other nations has indeed kept the German ships, naval and commercial, locked in port. But less than two hundred inexpensive submarines bid fair to sweep the seas of all merchant ships—neutral as well as British unless by feverish building the nations can build ships faster than submarines can sink them. Huge navies may henceforth be unknown.

The submarine has been the David of the war. It is a pity that its courage and efficiency have been exerted mainly in the wrong cause and that the missiles from its sling have felled the wrong Goliath.

Aircraft and submarine! It is still on the cards that when the definitive history of the war shall be written, its outcome may be ascribed to one or the other of these novel weapons—the creation of American inventive genius.

ALSO FROM LEONAUR
AVAILABLE IN SOFTCOVER OR HARDCOVER WITH DUST JACKET

IRON TIMES WITH THE GUARDS *by An O. E. (G. P. A. Fildes)*—The Experiences of an Officer of the Coldstream Guards on the Western Front During the First World War.

THE GREAT WAR IN THE MIDDLE EAST: 1 *by W. T. Massey*—The Desert Campaigns & How Jerusalem Was Won---two classic accounts in one volume.

THE GREAT WAR IN THE MIDDLE EAST: 2 *by W. T. Massey*—Allenby's Final Triumph.

SMITH-DORRIEN *by Horace Smith-Dorrien*—Isandlwhana to the Great War.

1914 *by Sir John French*—The Early Campaigns of the Great War by the British Commander.

GRENADIER *by E. R. M. Fryer*—The Recollections of an Officer of the Grenadier Guards throughout the Great War on the Western Front.

BATTLE, CAPTURE & ESCAPE *by George Pearson*—The Experiences of a Canadian Light Infantryman During the Great War.

DIGGERS AT WAR *by R. Hugh Knyvett & G. P. Cuttriss*—"Over There" With the Australians by R. Hugh Knyvett and Over the Top With the Third Australian Division by G. P. Cuttriss. Accounts of Australians During the Great War in the Middle East, at Gallipoli and on the Western Front.

HEAVY FIGHTING BEFORE US *by George Brenton Laurie*—The Letters of an Officer of the Royal Irish Rifles on the Western Front During the Great War.

THE CAMELIERS *by Oliver Hogue*—A Classic Account of the Australians of the Imperial Camel Corps During the First World War in the Middle East.

RED DUST *by Donald Black*—A Classic Account of Australian Light Horsemen in Palestine During the First World War.

THE LEAN, BROWN MEN *by Angus Buchanan*—Experiences in East Africa During the Great War with the 25th Royal Fusiliers—the Legion of Frontiersmen.

THE NIGERIAN REGIMENT IN EAST AFRICA *by W. D. Downes*—On Campaign During the Great War 1916-1918.

THE 'DIE-HARDS' IN SIBERIA *by John Ward*—With the Middlesex Regiment Against the Bolsheviks 1918-19.

AVAILABLE ONLINE AT **www.leonaur.com**
AND FROM ALL GOOD BOOK STORES

ALSO FROM LEONAUR
AVAILABLE IN SOFTCOVER OR HARDCOVER WITH DUST JACKET

FARAWAY CAMPAIGN *by F. James*—Experiences of an Indian Army Cavalry Officer in Persia & Russia During the Great War.

REVOLT IN THE DESERT *by T. E. Lawrence*—An account of the experiences of one remarkable British officer's war from his own perspective.

MACHINE-GUN SQUADRON *by A. M. G.*—The 20th Machine Gunners from British Yeomanry Regiments in the Middle East Campaign of the First World War.

A GUNNER'S CRUSADE *by Antony Bluett*—The Campaign in the Desert, Palestine & Syria as Experienced by the Honourable Artillery Company During the Great War .

DESPATCH RIDER *by W. H. L. Watson*—The Experiences of a British Army Motorcycle Despatch Rider During the Opening Battles of the Great War in Europe.

TIGERS ALONG THE TIGRIS *by E. J. Thompson*—The Leicestershire Regiment in Mesopotamia During the First World War.

HEARTS & DRAGONS *by Charles R. M. F. Crutwell*—The 4th Royal Berkshire Regiment in France and Italy During the Great War, 1914-1918.

INFANTRY BRIGADE: 1914 *by John Ward*—The Diary of a Commander of the 15th Infantry Brigade, 5th Division, British Army, During the Retreat from Mons.

DOING OUR 'BIT' *by Ian Hay*—Two Classic Accounts of the Men of Kitchener's 'New Army' During the Great War including *The First 100,000* & *All In It*.

AN EYE IN THE STORM *by Arthur Ruhl*—An American War Correspondent's Experiences of the First World War from the Western Front to Gallipoli-and Beyond.

STAND & FALL *by Joe Cassells*—With the Middlesex Regiment Against the Bolsheviks 1918-19.

RIFLEMAN MACGILL'S WAR *by Patrick MacGill*—A Soldier of the London Irish During the Great War in Europe including *The Amateur Army*, *The Red Horizon* & *The Great Push*.

WITH THE GUNS *by C. A. Rose & Hugh Dalton*—Two First Hand Accounts of British Gunners at War in Europe During World War 1- Three Years in France with the Guns and With the British Guns in Italy.

THE BUSH WAR DOCTOR *by Robert V. Dolbey*—The Experiences of a British Army Doctor During the East African Campaign of the First World War.

AVAILABLE ONLINE AT **www.leonaur.com**
AND FROM ALL GOOD BOOK STORES

ALSO FROM LEONAUR
AVAILABLE IN SOFTCOVER OR HARDCOVER WITH DUST JACKET

THE 9TH—THE KING'S (LIVERPOOL REGIMENT) IN THE GREAT WAR 1914 - 1918 by *Enos H. G. Roberts*—Mersey to mud—war and Liverpool men.

THE GAMBARDIER by *Mark Severn*—The experiences of a battery of Heavy artillery on the Western Front during the First World War.

FROM MESSINES TO THIRD YPRES by *Thomas Floyd*—A personal account of the First World War on the Western front by a 2/5th Lancashire Fusilier.

THE IRISH GUARDS IN THE GREAT WAR - VOLUME 1 by *Rudyard Kipling*—Edited and Compiled from Their Diaries and Papers—The First Battalion.

THE IRISH GUARDS IN THE GREAT WAR - VOLUME 1 by *Rudyard Kipling*—Edited and Compiled from Their Diaries and Papers—The Second Battalion.

ARMOURED CARS IN EDEN by *K. Roosevelt*—An American President's son serving in Rolls Royce armoured cars with the British in Mesopatamia & with the American Artillery in France during the First World War.

CHASSEUR OF 1914 by *Marcel Dupont*—Experiences of the twilight of the French Light Cavalry by a young officer during the early battles of the great war in Europe.

TROOP HORSE & TRENCH by *R.A. Lloyd*—The experiences of a British Lifeguardsman of the household cavalry fighting on the western front during the First World War 1914-18.

THE EAST AFRICAN MOUNTED RIFLES by *C.J. Wilson*—Experiences of the campaign in the East African bush during the First World War.

THE LONG PATROL by *George Berrie*—A Novel of Light Horsemen from Gallipoli to the Palestine campaign of the First World War.

THE FIGHTING CAMELIERS by *Frank Reid*—The exploits of the Imperial Camel Corps in the desert and Palestine campaigns of the First World War.

STEEL CHARIOTS IN THE DESERT by *S. C. Rolls*—The first world war experiences of a Rolls Royce armoured car driver with the Duke of Westminster in Libya and in Arabia with T.E. Lawrence.

WITH THE IMPERIAL CAMEL CORPS IN THE GREAT WAR by *Geoffrey Inchbald*—The story of a serving officer with the British 2nd battalion against the Senussi and during the Palestine campaign.

AVAILABLE ONLINE AT **www.leonaur.com**
AND FROM ALL GOOD BOOK STORES

www.ingramcontent.com/pod-product-compliance
Lightning Source LLC
Chambersburg PA
CBHW030216170426
43201CB00006B/105